Also by Roger Ellman:

* *THE ORIGIN AND ITS MEANING*

> *ON THE ORIGIN OF THE UNIVERSE*
> *AND ITS MECHANICS,*
> *THE MECHANISM AND ORIGIN*
> *OF INTELLIGENCE,*
> *AND THE IMPLICATIONS*
> *FOR THE INDIVIDUAL AND SOCIETY*

* *GRAVITICS*

> *THE PHYSICS OF THE BEHAVIOR AND CONTROL OF*
> *GRAVITATION*

* *THE PHILOSOPHIC PRINCIPLES OF RATIONAL BEING*

> *ANALYSIS AND UNDERSTANDING OF*
> *REALITY, TRUTH, GOODNESS, JUSTICE, VIRTUE, BEAUTY,*
> *HAPPINESS, LOVE, HUMAN NATURE, SOCIETY,*
> *GOVERNMENT, EDUCATION, DETERMINISM, FREE WILL, AND*
> *DEATH*

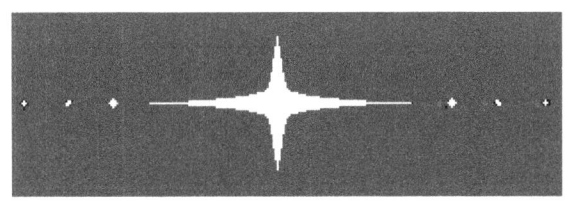

$$u(t) = U_c \cdot [1 - Cos(2\pi f t)] \cdot \varepsilon^{-t/\tau}$$

ON THE NATURE OF MATTER

THE ORIGIN OF THE UNIVERSE CREATED MATTER
FUNDAMENTALLY WAVE IN NATURE, NOT PARTICULATE

THE ORIGIN OF MATTER: ITS CAUSE;
THE STRUCTURE OF MATTER: ITS FORM;
MATTER'S INTERACTIONS: COULOMB, AMPERE, NEWTON,
MATTER WAVES, ATOMIC ORBITAL ELECTRONS,
GRAVITATION;
MATTER AND ATOMS;
APPLICATIONS

ROGER ELLMAN

Cataloging Data

Ellman, Roger (1932-   )

On the Nature of Matter

The origin of matter and its resulting characteristics, behavior, and applications.

Library of Congress Control Number: 2017918355

Published by: The-Origin Foundation, Inc.,
                 1401 Fountaingrove Pkwy.
                 Santa Rosa, CA 95403, USA

                 707-537-0257

                 http://www.The-Origin.org

ISBN 1981856978

CONTENTS

i

$\longrightarrow$

ABOUT THE AUTHOR

The-Origin Foundation, Inc. is a non-profit organization founded to foster independent scientific, mathematical, and philosophical research.

The author of "On the Nature of Matter", Roger Ellman, is the General Director of the foundation.

Roger Ellman has published over fifty professional papers on topics ranging from physics, cosmology, and astrophysics to artificial intelligence and mathematics.

He has presented some of his papers to conferences of / at:

The American Physical Society [APS], .
The American Society for the Advancement of Science,
Cambridge University, United Kingdom
The Library of Alexandria, Egypt
The Russian Academy of Natural Sciences, St Petersburg
The Hungarian Academy of Sciences, Budapest
A Science Conference in Shang Hai, China

He is author of three books in addition to the present "On the Nature of Matter".

His education includes graduate studies at Stanford University after graduating from West Point, the United States Military Academy.

PREFACE

In order to correctly understand the nature of matter it is necessary to consider all of the applicable sources of information and data. There are two such sources:

- The behavior of matter in its various encountered circumstances, and

- The origin of matter – how and from what it came to be.

The behavior of matter has been thoroughly investigated over the years and is codified in what we may refer to as 20th Century physics. That is the starting point of this present work, the various "Laws", "particles", "forces" and so forth that are the current generally accepted understandings of how the material world functions physically.

Until the present the origin of matter, its source, has not been addressed and that omission has resulted in a major error in the understanding of the nature of matter – the incorrect solution to the problem of the wave nature of matter versus its particle nature.

In the history of the physics of particles their wave aspect appeared significantly after their particle aspects had been well developed. Furthermore, although the wavelength aspect of matter waves became readily developed and experimentally confirmed the frequency aspect of matter waves could not then be successfully treated.

The late appearance of matter waves and the failure to treat their frequency resulted in the dominant success and acceptance of the general particle nature of matter.

That success has been so dominant that particle solutions to new phenomena are consistently proposed and the designation of new particles with names ending in "on", an imitating of pro<u>ton</u>s, elect<u>ron</u>s, and neut<u>ron</u>s, has become frequent.

It is unfortunate that a number of problems with current physics theory and a number of new avenues for physics investigation have been ignored. For example:

- Matter waves.
- Why are the stable atomic electron orbits the only stable ones ?
- How does one charge exert a force on another distant charge ? By electric "field" ? That is merely an assignment of a vector [a magnitude and direction] to each point in a subset of space without a supporting mechanism or cause.
- The same for the magnetic "field".
- How do the Lorentz Contractions occur since they are actual changes not mere observational differences ?
- What enforces the orbital electron structure as defined in terms of four "quantum numbers" n, l, m_l, and m_s.
- *etc.*

The present work resolves all of these issues and, making use of both of the above named applicable sources and of the developing of a solution to the matter wave frequency problem, presents, describes and advocates that all matter is wave in nature, oscillatory in nature.

Physics scientists have yet to offer explanation of the cause of the "Big Bang" – how and why the universe first came into existence.

But, the nature of that first cause dictated the nature of the universe to which it gave birth; it dictated the nature of its, our, on-going physics.

This book presents the only possible physics cause of that birth of the universe and presents from it the resulting physics of our world.

It validates that cause by directly, mathematically showing how the fundamental laws of physics, heretofore obtained only empirically, arise from that birth of the universe.

SECTION 1

The Origin of Matter: Its Cause

INTRODUCTION

In order to correctly understand the nature of matter it is necessary to consider all of the applicable sources of information and data. There are two such sources:

- The behavior of matter in its various encountered circumstances, and

- The origin of matter – how and from what it came to be.

Causality or mechanism is apparent from observation and experience which show that every thing and every event has a cause, and that those causes are themselves the results of precedent causes, and *ad infinitum*. Defining and comprehending the causality or mechanism operating to produce any contended or proposed scientific truth is essential to authenticating or validating that truth.

Until the present the science community has addressed only the first of those two with regard to the nature of matter and the omission of the second has resulted in a major error in the understanding of the nature of matter – the incorrect solution to the problem of the wave nature of matter versus its particle nature.

HOW THE UNIVERSE'S MATTER CAME TO BE

We are confronted with an apparently insuperable problem. Before the universe there was nothing, absolute nothing. That is the starting point because it naturally occurs; it is the only starting point that requires no cause, no explanation nor justification for its existence. But, that starting point has two impediments to the universe, or anything, coming into existence from it. First is the problem of change from nothing to something without, at least initially, an infinite rate of change, which is impossible. Second is the problem of change from nothing to something without violating conservation, which must be maintained.

The analysis would appear to end at that point, end with the declaration that obviously there cannot be a universe and there is no universe. Except, of course, that we and the universe we inhabit clearly exist at least enough for us to investigate it. Therefore, a solution to the insuperable problem exists. That solution is as follows.

1 - THE PROBLEM OF INFINITE RATE OF CHANGE

To avoid a material infinity the rate of change at the moment of the change must have been finite. Rather than an instantaneous jump from nothing to something, no matter how small or "negligible" that something might have been, there had to be a gradual transition at a finite rate of change. Further, the rate of change of that rate of change, the change's second derivative, at that moment had to have been finite, and so on *ad infinitum* for all of the further derivatives.

That requirement means that the form of the change had to have been either a natural exponential or some form of sinusoid. That develops as follows, in which the sought form of the change will be the function $U(t)$ [the "U" for universe, of course].

To illustrate the problem consider the function

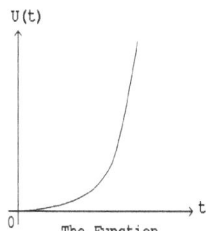
The Function

$(1-1)$ $\quad U(t) = 0 \qquad t < 0$
$\qquad\quad U(t) = t^2 \qquad t = 0 \text{ and } t > 0$

as a theoretical candidate for $U(t)$ at the beginning of the universe, which function is graphically depicted at the right.

Its first derivative, also depicted graphically to the right, is

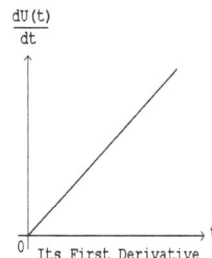
Its First Derivative

Figure 1-1a

$(1-2)$ $\quad \dfrac{dU(t)}{dt} = 0 \qquad t < 0$

$\qquad\quad \dfrac{dU(t)}{dt} = 2 \cdot t \quad t > 0$

and is unstated for $t=0$ because $dU(t)/dt$ is not smooth there even though $U(t)$ "looks" smooth there.

Now, the second derivative depicted graphically to the right

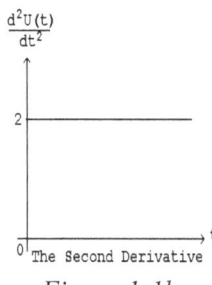
The Second Derivative

$(1-3)$ $\quad \dfrac{d^2U(t)}{dt^2} = 0 \qquad t < 0$

$\qquad\quad \dfrac{d^2U(t)}{dt^2} = 2 \qquad t > 0$

is clearly discontinuous at $t=0$, the instant of the beginning of the universe, where it instantaneously jumps from 0 to 2 as depicted to the right.

Figure 1-1b

The third derivative, which is the rate of change of the second derivative must be infinite at $t=0$ to produce the instantaneous jump from 0 to 2. Clearly, that cannot have happened in the real universe. It is such a condition which is unacceptable in a candidate function for $U(t)$ at the beginning of the universe.

The only way to avoid that condition of an infinite derivative somewhere along the line of successive further derivatives is to have a function with an endless family of finite, non-zero derivatives; that is, some derivatives may be zero at $t=0$ but there must always be further non-zero higher derivatives, which requires that the functional form of every derivative must be non-zero.

One can conceive theoretically of the idea of a function for which all derivatives are non-zero and no two are alike (in a general sense analogous to the pattern of digits in an irrational number), but it is not likely that such a function can exist. In any case the more certain and more simple way to achieve all non-zero derivatives is a repeating derivative function, the two simplest examples of which are as below.

(1-4)
$$\frac{dU(t)}{dt} = \pm\ U(t) \quad \text{[First derivative = the original function]}$$

(1-5)
$$\frac{d^2U(t)}{dt^2} = \pm\ U(t) \quad \text{[Second derivative = the original function]}$$

a. *Analysis of Repeating Derivative Functions*

Case (a): *Functions Satisfying Equation 1-4*

The function meeting this requirement is the natural exponential, ε^t.

(1-6)
$$\varepsilon^t = 1 + t + \frac{t^2}{2!} + \frac{t^3}{3!} + \cdots$$

Taking the first derivative

(1-7)
$$\frac{d[\varepsilon^t]}{dt} = 0 + 1 + \frac{2t}{2!} + \frac{3t^2}{3!} + \cdots$$
$$= 1 + t + \frac{t^2}{2!} + \frac{t^3}{3!} + \cdots = \varepsilon^t$$

so that the original function results as is required by equation *1-4*.

That is the prime case of a function that satisfies the requirement of all derivatives existing in functional form. In general those of this case are as equation *1-8*.

(1-8) $U(t) = A\cdot\varepsilon^t$

The function ε^t is not suitable for $U(t)$ at the beginning of the universe, however, because its value at $t=0$ is not zero. In fact it is zero only at $t = -\infty$. A function that might seem usable, however, would be

(1-9) $U(t) = 0 \qquad\qquad t < 0$ and $t = 0$

$\ \ U(t) = \varepsilon^t - 1 \qquad t > 0$

$$= t + \frac{t^2}{2!} + \frac{t^3}{3!} + \cdots$$

which does have zero value at $t=0$ and otherwise meets the derivatives requirement sufficiently.

Cases (b) – (e): *Functions Satisfying (1-5)*

Turning to functions that meet the requirement that the second derivative equal the original function per equation *1-5* there are four such functions.

(1-10)

$$\text{Case (b):}\quad U(t) = 1 + \frac{t^2}{2!} + \frac{t^4}{4!} + \cdots$$

3

(1-11)

$$\text{Case (c):} \quad U(t) = 1 - \frac{t^2}{2!} + \frac{t^4}{4!} + \cdots$$

(1-12)

$$\text{Case (d):} \quad U(t) = t + \frac{t^3}{3!} + \frac{t^5}{5!} + \cdots$$

(1-13)

$$\text{Case (e):} \quad U(t) = t - \frac{t^3}{3!} + \frac{t^5}{5!} + \cdots$$

These five candidate functions can be described and summarized as their exponential equivalents as in Figure 1-2, below.

Case	Function	Name of Function	Candidate U(t)
(a)	ε^t	Natural exponential	$\varepsilon^t - 1$
(b)	$\dfrac{\varepsilon^t + \varepsilon^{-t}}{2}$	Hyperbolic cosine	$\text{Cosh}(t) - 1$
(c)	$\dfrac{\varepsilon^{i \cdot t} + \varepsilon^{-i \cdot t}}{2i}$	Cosine	$\text{Cos}(t) - 1$
(d)	$\dfrac{\varepsilon^t - \varepsilon^{-t}}{2}$	Hyperbolic sine	$\text{Sinh}(t)$
(e)	$\dfrac{\varepsilon^{i \cdot t} - \varepsilon^{-i \cdot t}}{2i}$	Sine	$\text{Sin}(t)$

Figure 1-2

The relationships in the table can be verified by substitution using the formula for ε^t as given in equation *1-6*, above. Cases *(b)* and *(c)* have the same problem that case *(a)* had, that the value of *U(t)* is not zero at *t=0*. Just as with case *(a)*, they would appear to become satisfactory if a constant, *1*, is subtracted from each of them.

These candidates all satisfactorily meet the requirement for a continuous family of derivatives so that the kind of unacceptable problem as encountered in the example of $U(t)=t^2$ at the beginning of this discussion is avoided. That is, all derivatives are finite. But, there are other requirements that the successful *U(t)* function must meet.

b. *Using the Remaining Criteria to Select U(t)*

Two other criteria must be met by the successful candidate function or functions:

- the function must not be open-ended, that is it cannot ever have an infinite amplitude, and

- the function must smoothly match the *U(t)=0* condition at *t=0*.

4

The first criterion eliminates cases *(a)*, *(b)* and *(d)* each of which goes to an infinite value of $U(t)$. To satisfy the second criterion the tangent to $U(t)$ at $t=0$ must be identical to the tangent to the function for $t < 0$, which is the horizontal t-axis. The condition is satisfied if the first derivative of $U(t)$ equals *zero* at $t=0$. Only cases *(b)* and *(c)* meet that requirement.

Therefore, the resulting form of $U(t)$, the only acceptable form, the only one that meets all of the requirements, is case *(c)*,

$(1-14)$ $U(t) = [Cos(t) - 1]$ $\qquad\qquad$ $t > 0$ and $t = 0$

$\qquad\qquad$ $U(t) = 0$ $\qquad\qquad\qquad\qquad$ $t < 0$.

which is identical in form to the more usual and convenient equation *1-15*.

$(1-15)$ $U(t) = U_0 \cdot [1 - Cos(2\pi \cdot f \cdot t)]$

in which an amplitude parameter, U_0, and a frequency parameter, f, have been added.

That the only possible form for the manner in which the universe began is a sinusoidal oscillatory form would seem to be very appropriate. Oscillations, waves, are ubiquitous in our universe from oceans, violin strings and pendulums to sound, light and electron orbits. That statement can also be validly inverted: Oscillations and waves are ubiquitous in our universe because the universe began from an initial such oscillatory form.

Every oscillation that we know in nature exhibits, and the very theory of oscillations in the abstract requires, that the oscillation consist of two aspects storing and exchanging the energy of the oscillation back and forth by means of a "flow". (With one aspect varying in oscillatory fashion then when that aspect decreases there must be some "place" for its energy to go, a place in which it is stored until it reappears in that aspect when it increases again. It cannot completely disappear or be lost because the oscillation would die. That "place" is the oscillation's second aspect and it obviously must vary in a manner related to the first aspect's variation, but with its energy storage in opposite phase.

A pendulum, for example, oscillates by the motion (flow) of its swinging mass between peak height in the gravitational field (potential energy) at each end of the swing and peak speed of motion (kinetic energy) at the mid-point between the ends of the swing. Then, what is the "flow" of the original oscillation at the start of the universe ? We do not know and likely will never know but we can give it a name, *Medium*, and we can investigate its characteristics and nature.

Such was the oscillation at the beginning of the universe except that at the first half cycle the energy was in only one form increasing from zero to its maximum. Then the second form began, similarly from zero to maximum, receiving and storing the energy of the first form as that gradually decreased in the second half cycle.

2 - *THE PROBLEM OF CONSERVATION – "SOMETHING FROM NOTHING"*

At this point, that is the universe having started from absolute nothing as an oscillation having the form of equation *1-15*, the maintaining of conservation, the avoiding of getting something from nothing, clearly could only happen in one manner:

There simultaneously had to have arisen an identical-in-form but opposite-in-amplitude oscillation so that the pair balanced out to the original net nothing, as in equation *1-16*.

(1-16) $U(t) = \pm U_0 \cdot [1 - Cos(2\pi \cdot f \cdot t)]$

There is no other way that violating the assured principle of conservation could have been avoided. The universe exists. It had to come into being from a prior nothing. That had to happen while avoiding an infinity of rate of change. Conservation had to be maintained. The universe began with the oscillation of equation *1-16*.

3. THE PROBLEM: WHY THAT OSCILLATION BEGAN AND WHAT IT WAS

a. Why That Beginning happened

A duration is the period of time that a particular state or set of conditions persists. The duration is terminated by a change, which change also initiates a new duration. In the universe change is ubiquitous. It is the constant and continuous stream of change that makes durations mensurable. Before the beginning of the universe a duration was in process even though it was not mensurable. The beginning of the universe was the first change ever and it terminated the original primal duration of absolute nothing.

The probability of the happening of such an event is extremely small. But the event was / is not impossible. Furthermore, in the absence of that event occurring there was an extremely large duration of opportunity in which that extremely small probability could operate. In the absence of the beginning the original duration would have been infinite and that infinite opportunity operated on by minute, but non-zero, probability results in absolute certainty. The beginning of the universe could not avoid eventually happening.

b. What That Beginning Oscillation Was

The starting point is the assumption that, when the primal nothing changed as a probabilistically inevitable interruption of what would otherwise have been an infinite duration of the primal nothing, the simplest or minimum conservation-maintaining interruption that could occur is what occurred. There are two reasons for this. Occam's Razor, calls for the simplest hypothesis as the most likely. More importantly, or perhaps the same thing, if an essentially spontaneous and extremely low probability event is to occur solely as an interruption of the duration of an otherwise absolute nothing, then very little interrupting event is needed; the barest minimum of something is sufficient to interrupt, to be a change in absolute nothing. There is no call, no reason for anything more. So, while the interruption could have been otherwise, it was probably as simple and minimum as possible.

Size or amount of time are of no meaning here because there is nothing to which they can be compared or by which they can be measured. Whatever amount of change occurred is what occurred. Whatever time it took, or went on for, whatever its oscillatory frequency was, is what happened. Twice as much or half as much have no meaning.

The following conclusions about the initial oscillatory $\pm U_0 \cdot [1 - Cos(2\pi \cdot f \cdot t)]$ form can now be reasonably obtained:

- clearly the universe of today must be an on-going evolved consequence
  of its beginning, of the initial oscillatory form;

- the frequency, f, of the sinusoidal oscillation was, and is, very large; and

- the nature of the change is one of concentration or density of the something that is oscillating.

- the oscillation was spherical, radially outward in all directions from its origin, because there was nothing to constrain it otherwise.

The frequency would have to be either very large or very small -- high enough so that it is not detected or noticed by us in every day life or so low that it appears to us as no change at all in our experience.

It has already been noted that the fact that the only possible form for the manner in which the universe began is a sinusoidal oscillatory form is very appropriate because oscillations, waves, are ubiquitous in our universe from oceans, violin strings and pendulums to sound, light and electron orbits. And it has been noted that that statement can be validly inverted: oscillations and waves are ubiquitous in our universe because the universe began from an initial such oscillatory form.

If the frequency of the initial oscillation were so small that it appears to us as no change at all it would completely eliminate oscillations playing any significant part in the behavior of the universe as we know it. Therefore, the frequency must have been very large, so rapid compared to our perception that we do not notice the oscillation at all.

The change can hardly be one of gross size if it is going on right now at high frequency as has just been concluded. One can conceive of the fundamental "substance", the "something" of the universe flashing into and out of existence from a zero to a maximum density or concentration in an oscillatory fashion at a rate so high that we neither detect nor notice it at all. But, it is not possible to entertain a concept of reality flashing from zero to full size, a size that includes ourselves and our environment, in such a fashion.

Actually, the reality that we know is not "flashing into and out of existence ...." Our reality is more the oscillation itself than what is oscillating and the continuing oscillation is our steady, constant reality.

> Thus the interruption that gave us our universe was the starting of an *oscillation* that was *spherical*, present to us at a very high frequency and of $\pm U_0 \cdot [1 - Cos(2\pi \cdot f \cdot t)]$ form, of the density, as the variation will be hereafter referred to, of the *Medium*, as what it is that is oscillating will be hereafter referred to.

All of the discussion so far must apply to the "negative" oscillation, $-U(t)$, exactly as to the "positive" oscillation $+U(t)$ because the exact same reasoning as for $+U(t)$ applies to $-U(t)$ and, after all, they are not distinguishable in the discussion. The terms "+" and "−" are merely terms of convenience for two equal form opposite magnitude unknown things. We probably tend to think of our universe as the "+", but that is meaningless and irrelevant. There can be no objective designation of $+U(t)$ and $-U(t)$, no way to identify one versus the other. Both had to appear and our universe cannot avoid being the evolved result of both.

The universe that we know and exist in is the combined integrated result of both $+U(t)$ and $-U(t)$. The "+" and "−" electric charges of our universe [in both matter as

7

for example in protons and electrons and in anti-matter as for example in negaprotons and positrons] must derive from that aspect of the beginning. (It is interesting to observe, also, that our universe being the integrated result of an initial beginning and its opposite relates to (presumably is the underlying cause of) the dialectical nature of reality, the ying and yang of oriental philosophy.)

The question of what the *Medium* is can only be answered in terms of its characteristics, what it does and how. Its characteristics are:

- a continuous entity, not a mass of "particles" nor anything having parts,

- simple and uniform throughout,

- of minimum tangibility or substantiality, not unlike the actuality of what we designate as "field" [electric, gravitational, etc].

4. THE PROBLEM: WHY DID THE EFFECTS OF EQUATION *1-16* NOT PROMPTLY CANCEL AND ON-GOING ABSOLUTE NOTHING RESUME ?

This is resolved in detail in Appendix C, *Why No Immediate Mutual Annihilation*. Briefly, the initial structure was so unstable that it promptly exploded in that which we refer to as the "Big Bang" before annhilation could occur.

5. THE PROBLEM: IT HAS BEEN THOUGHT THAT THE UNIVERSE HAD TO START AT A POINT. HOW COULD A POINT DELIVER A WHOLE UNIVERSE?

The sole reason for positing a point origin was to avoid an initial infinite rate of change. The gradualness of the $[1 - Cosine]$ form resolves the problem of avoiding an infinite rate of change so that a point origin is no longer required.

The Big Bang "event horizon" problem and its relation to the development of variety in the universe has led to the hypothesis that there was an initial brief period of extremely rapid expansion called "inflation". That hypothesis has no supporting cause nor mechanism except its role in meeting the "event horizon" problem.

But with the need for a point origin eliminated the origin can have started per equation $1-16$ at any size. There was no un-accounted-for period of "inflation". From estimates calculated of the number of particles in today's universe it has been determined that the initial, at the very first instant, the already "inflated"–size universe began. It was a highly concentrated volume of all of the mass and energy of the universe of about $40,000 \ km$ radius.

That size is in terms of today's sizes. For that event specific size is meaningless because there was nothing else to compare it to.

$$\longrightarrow$$

$\longrightarrow$

The Behavior of Matter: Its Form

Section 1, *The Origin of Matter: Its Cause* resolved the origin of the matter of the universe as follows.

The universe exists. It had to come into being from a prior nothing. That had to happen while avoiding an infinity of rate of change. Conservation had to be maintained. *Ergo* equation $1-16$.

$(1-16)$ $\qquad U(t) = \pm\ U_0 \cdot [1 - Cos(2\pi \cdot f \cdot t)]$

Thus the hypothesis is that the interruption that started our universe, the interruption of what would otherwise have been an infinite duration of the primordial absolute nothing, an interruption because an essentially infinite amount of opportunity operated on a non-zero though minute probability, was the starting of a matched pair of spherical oscillations:

- Present to us at a very high frequency,

- Of the general *[1 - Cosine]* form, and

- Together equal to the original nothing because of having
  matching amplitudes $+U_0$ and $-U_0$.

That analysis yielded an initial event, the origin oscillations, as in Figure 2-1. [All of the unavoidably planar depictions of the spherical oscillations are of the spherical phenomenon, interpretable as a radial versus time depiction.]

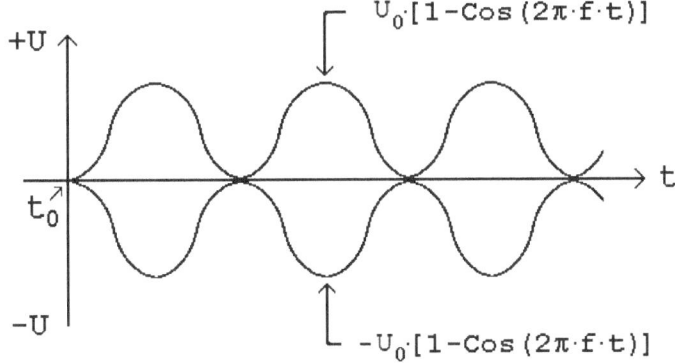

Figure 2-1

HOW THE ORIGINAL OSCILLATIONS BECAME THE UNIVERSE

Examination of the waveform of Figure 2-1 reveals two problems. One, that it is an immediate mutual annihilation, will be dealt with shortly below. Of concern now is that an infinite rate of change still remains; the envelope of the oscillation has an infinite rate of change at $t = t_0$ as can be seen in Figure 2-2, below, which displays the envelope.

11

Viewed in a mathematical or graphical sense without any consideration of the physical reality represented, the envelope discontinuity at $t=t_0$ is not a difficulty. The only quantity that actually exists and is varying is the overall $U(t)$. The envelope is merely our perception of a characteristic of the waveform. The actual varying quantity, per Figure 2-1, has no discontinuity at $t=t_0$

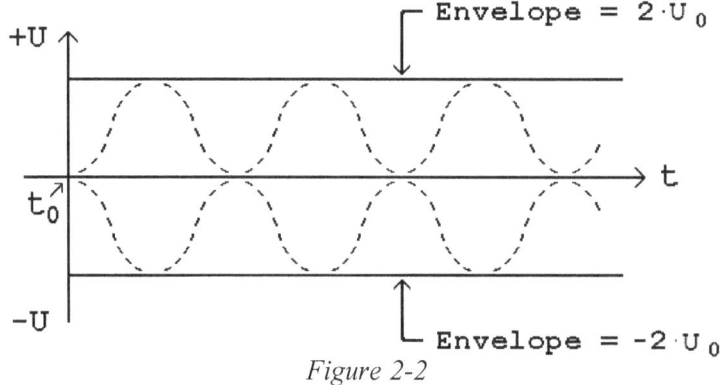

Figure 2-2

However, looking at the situation in a physical sense rather than purely mathematically, such oscillations as depicted in Figure 2-1 are all that there is to account for the effects which we call *energy, mass* and *charge*. Therefore, this *energy / mass / charge / oscillation* is something other than nothing. It is a physical reality that did not exist prior to the Origin. It can no more leap from zero to a finite non-zero amount than could the original $U(t)$ so leap.

That infinite rate of change in the amount of *energy / mass /charge* at $t=t_0$ is no more acceptable than was the infinite rate of change encountered in the original analysis of the beginning and it must be corrected by the same kind of reasoning as was then pursued: the envelope, also, had to originate as a *[1 - Cosine]* form of oscillation, which is the only form that avoids an infinite rate of change and matches the requirements of the situation.

That original envelope oscillation was at a lesser frequency than the original wave by the definition of a waveform envelope. If it were at a greater frequency then the roles (envelope and wave) would be reversed. If it were at the same frequency it would not act as an envelope and the infinity problem would remain. If we designate the envelope frequency as f_{env} and the frequency of the wave oscillation within the envelope as f_{wve} then the envelope would be of the following form.

(2-1) $U_{env} = [1 - \text{Cos}(2\pi \cdot f_{env} \cdot t)]$

The wave is, as before, of the form

(2-2) $U_{wve} = \pm U_0 \cdot [1 - \text{Cos}(2\pi \cdot f_{wve} \cdot t)]$

and the envelope modulating the wave is then

(2-3) $U(t) = [U_{env}] \cdot [U_{wve}]$

$= \pm U_0 \cdot [1 - \text{Cos}(2\pi \cdot f_{env} \cdot t)] \cdot [1 - \text{Cos}(2\pi \cdot f_{wve} \cdot t)]$.

That waveform appears in Figure 2-3.

However, the form of $U(t)$ of equation *2-3* and Figure 2-3 still does not resolve the problem of an infinite rate of change at t_0. The *[1 - Cosine]* envelope is itself an oscillation that begins at t_0 with a sudden step from zero to its full amplitude. Figure 2-3 shows the first *2* cycles of the envelope oscillation, which if only the envelope is considered, is a simple oscillation at the envelope frequency, even though visually, in the Figure, it is only the trace of the peaks of the overall complex oscillation.

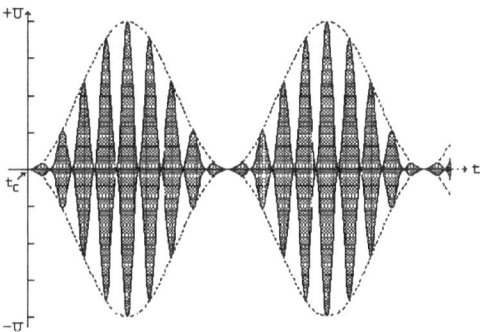

Figure 2-3

It is *energy / mass / charge* that begins suddenly in its full amount at t_0 just as, in Figure 2-1, the oscillation of equation *2-1* begins at t_0. Therefore, it is again necessary to introduce an envelope of *[1 - Cosine]* form to prevent the infinite rate of change at t_0 in the prior envelope. That correction will in turn require still another such correction and so *ad infinitum*. An (apparently at this point) infinite string of envelopes thus results as a necessity of the situation.

The resulting $U(t)$ then is

(2-4)

$$U(t) = \pm U_0 \cdot \prod_{i=1}^{i=\infty} \left[[1 - Cos(2\pi \cdot f_{env_i} \cdot t)] \right] \cdot \ \cdots$$

$$\cdots \ \cdot \left[1 - Cos(2\pi \cdot f_{wve} \cdot t) \right]$$

where the $\prod$ symbol (a large π, Greek "p") means the product of the indicated factors.

While an envelope frequency must be less than the frequency of the wave that it modulates so that the various f_{env} must be less than f_{wve}, each successive envelope may be at the same frequency, as the prior. The reason is as follows.

If each envelope frequency must be different then each must be at least slightly smaller than the prior. With an infinite set of envelopes and only the frequency range from slightly less than that of the wave down to slightly above zero being available each successive envelope could only be at an infinitesimally lower frequency than its predecessor in any case. Infinitesimally less is essentially the same as identical.

Then how did other than an infinite string of envelopes come about ?

13

Each additional envelope factor in equation *2-4* results in a higher frequency content in the overall expression. That is, as each envelope is added the expansion of the exponentiated cosines expression into a sum of individual frequency cosine terms becomes longer and acquires higher frequency terms. But, the oscillation could not have had an actual component at infinite frequency. The real universe original *U(t)* had an enormous set of envelopes but not an infinite set; they were "cut off" at some point.

The *Medium* of these oscillations being the only reality and, therefore, being what sets the limit on the speed of light with which we are familiar, the *Medium* also sets a limit on the highest frequency / lowest wavelength waves that can propagate. As a result the series of envelopes, of factors in equation *2-4*, was limited to some finite but quite large amount. (See Appendix B, *The Limitation of the Original Envelopes*).

This yields a revised *U(t)*, the original oscillation, the Cosmic Egg, as equation *2-5*, below. N_0 is the number of envelopes, all at the same frequency, f_{env}.

(2-5) $$U(t) = \pm U_0 \cdot \left[1 - Cos\left[2 \cdot \pi \cdot f_{env} \cdot t\right]\right]^{N_0} \cdot \left[1 - Cos\left[2 \cdot \pi \cdot f_{wve} \cdot t\right]\right]$$

The waveform *[1 - Cos(x)]^n* converges to an increasingly narrower peak as *n* increases, Figure 2-4, below. For very large *n*, that is very large N_0 of equation *2-5*, the converging of the waveform into a single narrow peak proceeds to a momentary "spike" per cycle. Figure 2-5, below, shows the appearance of the waveform for extremely large *n*, that is for *n* = N_0 - what the waveform of the original "Cosmic Egg", the start of our universe, "looked like". (N_0 is found further below to be about 10^{84}.)

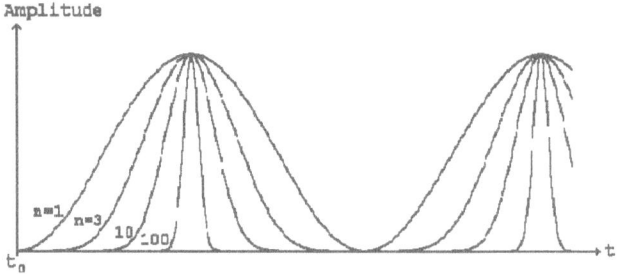

Figure 2-4 [1 - Cos(x)]^n For n = 1, 3, 10, 100

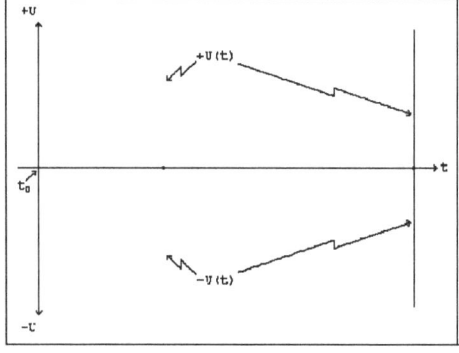

Figure 2-5
The U(t) "Cosmic Egg" WaveForm

14

This discussion of $U(t)$, the original oscillation the start of which was the start of the universe, has dealt so far only with the problems of the Origin, the problems of the transition from nothing to something. The something was, of course, the first instant of the entire universe. As such it must have contained in itself all of the *mass / energy / positive and negative charge* of the universe.

Figures 2-1, 2-3, and 2-5 all indicate that the original pair of oscillations, $+U$ and $-U$, should have immediately mutually annihilated, canceled out, reverted to the primal nothing. But, clearly that did not happen. The only explanation of that not happening is that each was unstable, so unstable that they exploded more immediately than they were able to mutually annihilate. They immediately proceeded to an immense explosion of energy and pieces of their oscillation, the event now called the "Big Bang". See Appendix C, *Why No Immediate Mutual Annihilation.*

In terms of the $U(t)$ as depicted in Figure 2-5, the so immediate explosive decay undoubtedly occurred after only a minute portion, an infinitesimal portion, of the very first cycle had passed. It had to have been long before the first "spike". In that sense the initial event was very small, tenuous, hardly more than nothing because the instantaneous amplitude of $U(t)$ at that moment (the height of the curve above zero at that moment long before the first "spike") was also infinitesimal. It was hardly more than, essentially zero.

In that sense, the way that the universe started at all becomes a little more comprehensible. To avoid an infinite rate of change there was essentially almost no difference between "nothing", on-going absolute nothing, and the first infinitesimal moment of the original $U(t)$, the original oscillation.

Yet, it contained the entire universe.

THE FORM OF MATTER AS GENERATED BY THE "BIG BANG"

What did the "Cosmic Egg" explode into ? It could only explode into pieces of what it was made of, pieces of *[1 – cosine]* form spherical oscillations, pieces like equation $1-16$, above.

Each oscillation is three-dimensional, thus spherical, because three dimensions is the minimum number that can involve space part of which is not its own boundary.

But, what did the "Cosmic Egg" explode into ? It primarily exploded into what we know our universe to mainly consist of: myriad protons - Hydrogen atom nuclei, and myriad electrons - maintaining overall charge neutrality with the protons, and the antimatter forms of both, negaprotons and positrons – maintaining conservation.

[Those might also be expected to have mutually annihilated but did not. Their survival rather than annihilation is analyzed in full in Appendix C, *Why No Immediate Mutual Annihilation.* Suffice it here to observe that each product piece was initially ejected radially outward at extreme velocity and energy, on paths slightly diverging, such that initially annihilations could not occur.]

Then, what was the nature, the form of those product pieces that the "Cosmic Egg" exploded into ? Because of the two frequencies of $U(t)$, f_{wve} and f_{env}, and that the explosion source was of two equal but opposite polarities, $+U_0$ and $-U_0$, the "Big Bang" resulted in myriad pieces of four different forms of *[1 – cosine]* form spherical oscillations , equations $2-6$.

(2-6) $U_{Form\ 1}(t) = +U_c \cdot [1 - Cos(2\pi \cdot f_{wve} \cdot t)]$ the proton

$U_{Form\ 2}(t) = -U_c \cdot [1 - Cos(2\pi \cdot f_{env} \cdot t)]$ the electron

$U_{Form\ 3}(t) = -U_c \cdot [1 - Cos(2\pi \cdot f_{wve} \cdot t)]$ the anti-proton

$U_{Form\ 4}(t) = +U_c \cdot [1 - Cos(2\pi \cdot f_{env} \cdot t)]$ the anti-electron

Each of those has a specific value of its mass. Per the data provided by NIST, the National Institute of Standards and Technology those masses are:

(2-6a)   ■ the proton and the antiproton $m_p = 1.672\ 621\ 898 \cdot 10^{-27}$ kg

■ the electron and the anti-electron $m_e = 9.109\ 383\ 56 \cdot 10^{-31}$ kg.

Using the mass-energy relationship, $m \cdot c^2 = h \cdot f$ the frequency, f, of those particles can be calculated. Those frequencies are:

(2-6b)   ■ the proton and anti-proton: $f_{wve} = 2.268,731,818 \cdot 10^{23}$ hz

■ the electron and anti-electron: $f_{env} = 1.235,589,965 \cdot 10^{20}$ hz.

Finally, the mass of those four fundamental particles having now been resolved, their electric charge remains. They all have the same magnitude of their oscillation, $|U_c|$, which by default is the magnitude of their electric charge. [U_c is the particle oscillation amplitude per equation 2-6. U_0 is the original pre-explosion oscillation amplitude.] The magnitude of the oscillation is in two opposite polarities; therefore clearly, where q is the fundamental electric charge per NIST, then:

(2-7) $q = 1.602,176,621 \times 10^{-19}$ c

$+U_c = +q$

$-U_c = -q$

Judging by its result, the "Cosmic Egg" was not unlike an immense atom, a very unstable immense atom [as are all of the atomic species of atomic number exceeding 83 which the cosmic egg would have immensely exceeded]. Its "Big Bang" was a kind of explosive nuclear radioactive decay ultimately ending in the myriad stable elements of today's Periodic Table plus those with half lives long enough to be in detectable quantities today. Such decays follow a chain:

- From a heavy and complex composition,

- To various multiple less heavy less complex product pieces,

· · · · ·

- Until they arrive at many multiple stable forms.

The vast majority of those resulting stable forms are the protons and electrons of the material world and their anti-particles. They are of the equation *2-6* form spherical oscillation, and will be referred to as *Spherical-Centers-of-Oscillation* or as *particles*

The rates of the decays are exponential, the decay [varying from some extremely rapid to some extremely slow] is described in terms of a "half life", the time it takes for half of the original material's decays to take place. Some of those "multiple less heavy less complex product pieces" having long half lives are present to us still today still decaying as what we term "radioactive" species.

The process of radioactive decay is treated in detail in Appendix A-3, *Radioactivity*. The atomic nucleus and various atomic species are treated in detail in Appendix A-2, *The Atomic Nuclei*.

The actions of the various stable atomic forms are primarily: electrostatic per Coulomb's Law, electromagnetic per Ampere's Law, and gravitational per Newton's Law. Those are treated in detail in the following sections; however, examination now of an aspect of gravitational behavior results in additional information on the behavior and form of matter as follows.

THE FLOW FROM THE SPHERICAL-CENTERS-OF-OSCILLATION

The Particle "Core"

Consider a small individual particle such as a proton. Newton's law of gravitation expressed in terms of m_{source} and $m_{acted-on}$ and with both sides of the equation divided by $m_{acted-on}$ is, of course,

$$(2-7) \qquad a_{grav} = G \cdot \left[\frac{m_{source}}{d^2} \right]$$

However, mass and energy are equivalent, so that [using c = light speed and h = Planck's constant] a mass, m, is proportional to a frequency, f, that is characteristic of that mass. That is

$$(2-8) \qquad m \cdot c^2 = h \cdot f \quad \text{or} \quad f = [c^2/h] \cdot m$$

so that the m_{source} of equation $2-7$ has a corresponding equivalent frequency, f_{source}.

That being the case, the gravitational acceleration, a_{grav}, can be expressed in terms of that frequency as the change, Δv, in the velocity, v, of the attracted mass per time period, T_{source}, of the oscillation at the corresponding frequency, f_{source}, as follows.

$$(2-9) \qquad a_{grav} = \Delta v \, / \, T_{source} = \Delta v \cdot f_{source}$$

It can then be reasoned using equation $2-9$ = equation $2-7$ as follows .

$$(2-10) \qquad a_{grav} = \Delta v \cdot f_{source} = G \cdot \left[\frac{m_{source}}{d^2} \right]$$

Equation $2-11$, below, is obtained by using that frequency is proportional to mass. With f_p and m_p as the proton frequency and mass then $f_{source} = [m_{source} \, / \, m_p] \cdot f_p$.

$$(2-11) \qquad \Delta v \cdot \left[\frac{m_{source}}{m_p} \right] \cdot f_p = G \cdot \left[\frac{m_{source}}{d^2} \right]$$

Rearranging and canceling m_{source} on both sides of the equation,

$$(2-12) \qquad \Delta v = \frac{G \cdot m_p}{d^2 \cdot f_p} \quad \text{per cycle of } f_{source}.$$

Then substituting, per equation $2-8$, $m_p = [h \cdot f_p] \, / \, c^2$,

(2-13)
$$\Delta v = \left[\frac{G}{d^2 \cdot f_p}\right] \cdot \left[\frac{h \cdot f_p}{c^2}\right]$$

$$= \frac{G \cdot h}{d^2 \cdot c^2} \quad \text{per cycle of } f_{source}.$$

The Planck Length, l_P, is defined as

(2-14)
$$l_P \equiv \left[\frac{h \cdot G}{2\pi \cdot c^3}\right]^{\frac{1}{2}} \quad \text{so that} \quad G = \left[\frac{2\pi \cdot c^3 \cdot l_p^2}{h}\right]$$

Substituting G as a function of the Planck Length from equation *2-14* into G as it is in equation *2-13*, the following is obtained.

(2-15)
$$\Delta v = \left[\frac{2\pi \cdot c^3 \cdot l_p^2}{h}\right] \cdot \left[\frac{h}{d^2 \cdot c^2}\right]$$

$$= c \cdot \frac{2\pi \cdot l_p^2}{d^2} \quad \text{per cycle of } f_{source}.$$

This result states that:
- the velocity change due to gravitation, Δv,
- per cycle of the attracting mass's equivalent frequency, $f_{source,}$
  which quantity, $\Delta v \cdot f_{source}$, is the gravitational acceleration, a_{grav},
- is a specific fraction of the speed of light, c, namely the ratio of:
  - 2π times the Planck Length squared, $2\pi \cdot l_P^2$, to
  - the squared separation distance of the masses, d^2.

That squared ratio is, of course, the usual inverse square behavior.

This also means that at distance $d = \sqrt{2\pi} \cdot l_P$ from the center of the source, attracting mass, the acceleration, Δv, per cycle of that attracting mass's equivalent frequency, f_{source}, is equal to the full speed of light, c, the most that it is possible to be. In other words, at that [quite close] distance from the source mass the maximum possible gravitational acceleration occurs. That is the significance, the physical meaning, of l_P or, rather, of $\sqrt{2\pi} \cdot l_P$.

The physical significance of $\sqrt{2\pi} \cdot l_P$ is that it sets a limit on the minimum separation distance in gravitational interactions and it implies that a "core" of that radius is at the center of fundamental particles having rest mass. That is, equation *2-15* clearly implies that it is not possible for a particle having rest mass to be approached closer than that distance.

That physical significance of $\sqrt{2\pi} \cdot l_P$, is so fundamental to gravitation and apparently to particle structure, that it more truly represents a fundamental constant than does l_P. For those reasons that length should replace l_P as a fundamental constant of nature as follows.

(2-16) The <u>fundamental distance constant</u>, δ

$$\delta^2 \equiv 2\pi \cdot l_P{}^2$$

$$\delta = 4.051,34 \times 10^{-35} \text{ meters}$$

Equation *2-15* then becomes equation *2-17*.

(2-17) $\quad \Delta v = c \cdot \dfrac{\delta^2}{d^2} \quad$ per cycle of f_{source}

a quite pure and precise statement of gravitation: that gravitation is a function of the speed of light, *c*, and the inverse square law, in the context of the oscillation frequency, *f_source*, corresponding to the attract**ing**, source body's mass.

It makes clear that an oscillation is an integral part of gravitation as should be the case because gravitation is an action between particles having mass, which are the just-developed *Spherical-Centers-of-Oscillation* products, equation *2-16,* of the "Big Bang". See Section 7, *The Action of Matter - Gravitation*.

The Particle Core's Propagated Outward Flow

Each gravitationally attract**ing** *Spherical-Center-of-Oscillation* must tell each gravitationally attract**ed** particle its "message": the direction from the attract**ed** particle to the attract**ing** one and the magnitude of the attract**ing** particle's gravitational attraction. That task is assigned by contemporary physics' theory to a *gravitational field*, a vector field that is an assignment of a direction of action and its magnitude to each point in a region of space.

However, that designation of the field, while facilitating the description of the action fails to explain the cause, the mechanism of the field and thus fails to explain or account for the action at issue. It also fails to account for the time delay due to the limitation of the speed of light that must exist between a change at the attract**ing** particle and its effect at the attract**ed** particle.

Something flowing is required, something flowing at the speed of light, continuously, carrying the direction and magnitude information, spherically outward, from every gravitating *Spherical-Center-of-Oscillation* to every other *Spherical-Center-of-Oscillation*.

Furthermore, the necessity for gravitation that an oscillation and its frequency are closely involved in the effect [equations *2-15* and *2-17*] and therefore in what is communicated by the flow, means that the <u>flow itself is oscillatory</u> corresponding to and generated by its oscillatory source, the *Spherical-Center-of-Oscillation*.

For such a flow to persist there must be a supply of that outward flowing substance in every particle. And, for that flow to have persisted the billions of years since the "Big Bang" that "supply" must be an extremely concentrated reservoir of that which flows outward [concentrated relative to the outward flow].

Having now just determined:

- That δ sets a limit on the minimum separation distance in gravitational interactions and therefore that a "core" of that radius is at the center of fundamental particles, and

- That an extremely concentrated reservoir supply of that which is flowing outward is required at the center of all particles to support the billions of years of their outward flow;

Therefore:

- The reservoir is the spherical "core" of radius δ at the center of all particles;
- That it is impenetrable is because of its immense density concentration [billions of years worth of flow of the flow substance [*Medium*] in the minute (δ = 4.05134 $\times$ 10^{-35} meters radius spherical core) of every particle having rest mass], and.
- The *Spherical-Center-of-Oscillation* is a spherical oscillation of that immensely concentrated flow substance, *Medium*.

Then, what "contains" that core's supply or why doesn't it all just quickly "slosh" out and be gone ? The answer is that it is trying to do just that, to "slosh" out, as hard as it can. It cannot help propagating outward because it has no container. But it can only propagate outward at the limiting rate determined by its surface area, $4 \cdot \pi \cdot \delta^2$ and the fastest speed possible for flow, the speed of light, c. Thus is the *Propagated Outward Flow*.

The Speed of the Flow – The Speed of Light

Every oscillation that we know in nature exhibits, and the very theory of oscillations in the abstract requires, that the oscillation consist of two aspects of the substance which is oscillating [e.g. pendulum position and velocity or electric potential and current] storing and exchanging back and forth the energy of the oscillation. With one aspect varying in oscillatory fashion then when that aspect decreases there must be some "place" for its energy to go, a place in which it is stored until it reappears in that aspect when it increases again. It cannot completely disappear or be lost because the oscillation would die. That "place" is the oscillation's second aspect and it obviously must vary in a manner related to the first aspect's variation with its energy storage in opposite phase.

The matter of the universe is largely a mass of particles each a spherical *[1 - Cosine]* form oscillation propagating outward.

Like electric inductance and capacitance determining the speed of propagation along a transmission line, μ_0 and ε_0 determine the speed of the *[1 - Cosine]* form oscillation propagation by setting the two aspects of the oscillation in which they are involved, the aspects between which the oscillation energy exchanges back and forth.

But, when the original oscillation came into existence it did so in absolute nothing. There was no "free space" with μ_0 and ε_0. There was nothing but the original oscillation. And, after the immediate explosion into all of the particles of the universe, each of those particles was sending its *Propagated Outward Flow* into nothing, into emptiness.

Where did the *Propagated Outward Flow*'s μ_0 and ε_0 come from? The only thing they could have come from was the original oscillation. There is no other possible source because everything else was absolute nothing, "the zero of existence". The μ_0 and ε_0 are inherent in the substance of the oscillation, which means, μ_0 and ε_0 are also inherent in the outward propagation. Each particle's *Propagated Outward Flow* contains its own μ_0 and ε_0.

Having established the supply of *Medium* [flow substance] and its on-going *Propagated Outward Flow* serving the role of gravitational field [see Section 7, *The Action of Matter: Gravitation*] as a property of every particle exhibiting rest mass, the question arises, "What of the electric field, much stronger than gravitation and co-present with gravitational field whenever the gravitating particle has electric charge ?"

Just as is the case for gravitation, every particle having electric charge must tell its similar "message" to every other such particle [see Section 3, *The Action of Matter: The Electrostatic Effect - Coulomb's Law*]. That requires something flowing outward at the speed of light continuously, carrying the direction and magnitude information, spherically outward, from every electrostatic *Spherical-Center-of-Oscillation* to every other *Spherical-Center-of-Oscillation*. That flow-communication is the electric field, an active process not a static state.

The theory of an *electric field*, just as with that of a *gravitational field*, above, while facilitating the description of the action fails to explain the cause, the mechanism of the field and thus fails to explain or account for the action at issue. It also fails to account for the time delay due to the limitation of the speed of light that must exist between a change at the attract**ing** particle and its effect at the attract**ed** particle

Two such simultaneous flows, gravitational and electric, and two supporting reservoirs supplying the flows, is clearly untenable. There can only be one reservoir in each particle's "core" and one resulting *Propagated Outward Flow* producing both the gravitational action and the electric action if for no other reason than because two supply reservoirs would mutually interfere with a spherically outward flow of each.

The one, single, universal flow functions as follows. As developed fully in Section 3, *The Action of Matter: The Electrostatic Effect – Coulomb's Law*, that effect is due to impulses delivered by the source *Propagated Outward Flow* on the encountered *Spherical-Centers-of-Oscillation*. As developed fully in Section 7, *The Action of Matter: Gravitation*, that effect is due to the arriving flow's inverse square reduced μ and ε adding to the outgoing encountered flow's full magnitude μ and ε slowing that encountered flow.

SUMMARY FOR SECTION 2 – THE BEHAVIOR OF MATTER: ITS FORM

The form of matter is not that of the "particles" of classical modern physic's Standard Model. Rather the form of matter is:

- *Spherical-Centers-of-Oscillation*, spherical oscillations of *[1 - Cosine]* form, equation 2-6;

- Propagating spherically outward a continuous oscillatory *Propagated Outward Flow* of *Medium* in *[1 - Cosine]* form, according to its source *Spherical-Center-of-Oscillation* magnitude, sign, and frequency;

- The speed of the *Propagated Outward Flow*, c, being set by the net μ and ε in the *Medium* being propagated;

(2-18)
$$c = \frac{1}{\sqrt{\mu \cdot \varepsilon}}$$

The *Spherical-Center-of-Oscillation* consists of a central "core", a spherical volume of radius $\delta = 4.051,34 \times 10^{-35}$ meters that consists entirely of a high density concentration of the oscillating *Medium*, which propagates outward at an extremely low rate determined by the surface area of the "core" and the radial outward speed of flow of the propagated *Medium*, the speed of light, c.

THE GENERAL EXPONENTIAL DECAY OF THE UNIVERSE

Since the "Big Bang" the *Propagated Outward Flows* have been gradually depleting the original supply of *medium* in each *Spherical-Center-of-Oscillation*. That process, an original quantity gradually depleted by flow away of some of the remaining quantity is an exponential decay of the form equation *(2-19)*.

(2-19)

$$\upsilon(t) = \upsilon_0 \cdot \varepsilon^{-t/\tau} \quad - \upsilon \equiv \text{the amount of medium}$$

$$- \tau \equiv \text{the "time parameter"}$$

The value of τ is

(2-20) $\quad \tau = 3.57532 \cdot 10^{17}$ seconds

$\approx 11.3373 \cdot 10^{9}$ years

The decay is treated in detail in Appendix E, *The Universal Exponential Decay.*

$\longrightarrow$

$\longrightarrow$

SECTION 3

The Action of Matter: The Electrostatic Effect
Coulomb's Law

The fundamentals of what is known about the actions of electric charge can be summarized as follows.

- Electric charges exist in two different forms, termed the *sign* or *polarity* of the charge, "positive" or "negative".

- The charges exert a <u>force</u> on other electric charges.

- The <u>force</u> attracts or repels for charges of opposite or like signs.

- The <u>force</u> is inversely proportional to the square of the separation distance of the charges and directly proportional to the amounts of the charges.

- The effect extends throughout space.

- The charges only exist as a component effect of particles having mass which particles have been shown in the preceding sections to be *Spherical-Centers-of-Oscillation*.

The details of this behavior have been thoroughly worked out in terms of mathematics which describe the location, amount, and direction of the effect. The physical constants needed to give correct quantitative results have been well determined.

Each electrically forc**ing** particle *[Spherical-Center-of-Oscillation]* must communicate to each electrically forc**ed** particle *[Spherical-Center-of-Oscillation]* the direction from the forc**ing** particle to the forc**ed** one [for same signs repulsion], the direction from the forc**ed** particle to the forc**ing** one [for opposite signs attraction] and the magnitude and sign of the forc**ing** particle's charge. That task is assigned by contemporary physics' theory to an *electric field*, a vector field that is an assignment of a direction of action and its magnitude to each point in a region of space.

However, that designation of the field, while facilitating the description of the action fails to explain the cause, the mechanism of the field and thus fails to explain or account for the action at issue. It also fails to account for the time delay, due to the limitation of the speed of light, that must exist between a change at the forc**ing** particle and its effect at the forc**ed** particle

A flow, flowing at the speed of light, continuously, carrying the direction and magnitude information, spherically outward, from every electrically acting *Spherical-Center-of-Oscillation* to every other such *Spherical-Center-of-Oscillation*, from every charge to every other, is required. That *Propagated Outward Flow* was introduced and described in the preceding Section 2.

25

HOW THE CHARGES AND THEIR FLOW REPEL AND ATTRACT

The effect of an individual wave of that *Propagated Outward Flow* encountering another *Spherical-Center-of-Oscillation* is the delivery of a train of impulses to the center, Figure 3-1, each an amount of momentum. That is

(3-1)   impulse = force·time = mass·velocity = momentum

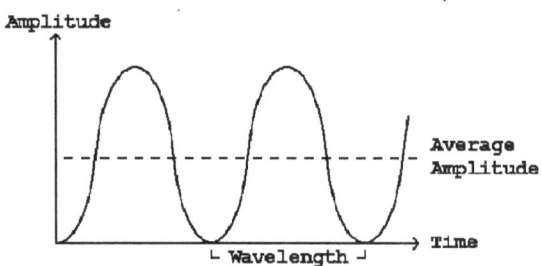

Figure 3-1
The +U Wave of the Propagated Outward Flow from a +U Spherical Center-of-Oscillation

The wave as it is propagated by its source *Spherical-Center-of-Oscillation*, carries potential impulse, "potential" because it is not realized in an effect until an encounter with another *Spherical-Center-of-Oscillation* occurs. The amount of potential impulse in the wave is, of course, proportional to the amplitude of the wave. It is that amplitude, which decreases as the square of the distance from the source *Spherical-Center-of-Oscillation* because it becomes spread over a greater area.

The overall stream of waves carries the potential impulse of one wave times the repetition rate, the frequency, of the waves. The potential status of the wave's impulse is exactly the same status as that of electric field (which it, in fact, is) where electric field is potential force and not realized as actual force until it interacts with another charge.

A *Spherical-Center-of-Oscillation* propagating a *+U* Wave *Propagated Outward Flow* experiences an equal *Spherical-Center-of-Oscillation* magnitude, opposite direction reaction to the radially outgoing train of impulses as if the *Spherical-Center-of-Oscillation* were under spherical compression, Figure 3-2. However, that is to no net effect because of its spherical symmetry.

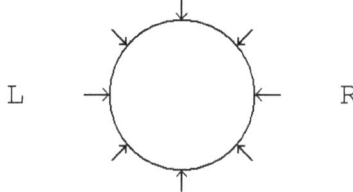

Figure 3-2
The +U Spherical-Center-of-Oscillation's Reaction Back On Itself by Its Outward Flow

The train of impulses of Figure 3-1 encountering the *Spherical-Center-of-Oscillation* of Figure 3-2 on its left side [L] adds additional momentum to the reaction directed to the right. That being now greater than the opposing reaction to the left on the right side [R], there is now a net momentum increment to the right, a repelling action of one positive charge on another.

26

If the *Spherical-Center-of-Oscillation* of Figure 3-2 were a -*U* center the effect would be reversed. The train of +*U* impulses of Figure 3-1 encountering the center of Figure 3-2 as a -*U* center on its left side [L] subtract from or cancel part of its reaction directed to the right. That being smaller than the opposing reaction to the left on the right side [R], there is a net momentum increment to the left. The effect is an attracting action of a positive charge on a negative one.

The effects and action are exactly analogous for the two other cases of a -*U* *Spherical-Center-of-Oscillation*'s train of -*U* impulses encountering another -*U* *Spherical-Center-of-Oscillation* or a +*U* one.

It is important to observe that the direction of momentum actions is the direction of the *Propagated Outward Flow* transmitting them whereas the sign or polarity +*U* or -*U* pertains back to the origin of the oscillations that started the "Big Bang" – a pair of exact opposites necessary to maintain conservation.

Having obtained from the *Spherical-Centers-of-Oscillation* and their *Propagated Outward Flows* the directions and polarities of Coulomb's Law it is now necessary to definitively quantify the action.

Newton's Law and Centers & Waves – "Responsiveness"

Newton's Second Law and as restated by inversion are:

(3-2)   Force = Mass · Acceleration

Acceleration Resulting = Force Applied × $1/_{Mass}$

which translates in terms of the waves of *Propagated Outward Flow*s and *Spherical-Centers-of-Oscillation*s into

$$(3\text{-}3) \quad \begin{bmatrix} \text{Acceleration} \\ \text{Resulting} \end{bmatrix} = \begin{bmatrix} \text{Wave} \\ \text{Impulse} \end{bmatrix} \cdot \begin{bmatrix} \text{Responsiveness} \\ \text{of the Center} \end{bmatrix}$$

or, more succinctly,

Acceleration = Wave × Responsiveness.

Of the total wave traveling outward from the source *Spherical-Center-of-Oscillation*, the only part that interacts with another *Spherical-Center-of-Oscillation* is the part intercepted by the encountered center. The *Spherical-Center-of-Oscillation* intercepting the larger portion of incoming wave receives the greater impulse, the greater momentum change. Thus center responsiveness depends on the encountered center's cross-section target for interception of *Propagated Outward Flow* waves.

(This analysis assumes that the part of the wave intercepted by the encountered center is a flat wave front. The non-plane wave case, for small separation distances, is in most cases of negligible effect except the slight "Lamb Shift" treated in Appendix A-1, *The Neutron*, Likewise, because δ, the encountered particle's core radius, is so minute the target can be deemed flat)

A *Spherical-Center-of-Oscillation* of smaller cross-section is of greater mass (lesser responsiveness). The encountered center being a spherical oscillation the cross-section is the area of a circle perpendicular to the direction of travel of the incoming wave front as it encounters the center. That area is <u>proportional </u>to π times the square of the center's wavelength .

This yields the first factor in *Spherical-Center-of-Oscillation* responsiveness,

(3-4)   Cross-section $\propto \ \pi \cdot \lambda_c^2 \ = \ K_{cs} \cdot \lambda_c^2$

(3-5) $\left[\begin{array}{c} \text{respon-} \\ \text{siveness} \end{array} \right] \propto$ [Factor 1]$\cdot$[Factor 2]$\cdot$[Factor 3]

                      $=$ [$K_{cs} \cdot \lambda_c^2$]$\cdot$["   ]$\cdot$["   ]

       where: K_{cs} = a constant for the proportionality
               λ_c = the encountered center oscillation wavelength

The incoming wave must be expressed in terms of "Wave Impulse per Unit Area" so that multiplied by the cross-section area at the encountered *Spherical-Center-of-Oscillation* the units of area are cancelled and the resulting quantity is wave impulse.

(3-6)

$$"Wave" \ = \ \frac{\text{Total Source Center Propagated Wave}}{\substack{\text{Total Spherical Area of Source Wave at} \\ \text{Distance Encountered Center is from Source}}}$$

         $=$ Wave Impulse per Unit Area

Factor 2 in the responsiveness, equation *3-5* is the <u>effective</u> amplitude of the *Spherical-Center-of-Oscillation*'s oscillation. A range of possible interactions can occur because the source and encountered center frequencies may differ. The extremes and mean of the range of encounters follow.

(1) Frequency$_{source}$ $\ll$ Frequency$_{encountered}$

The encountered center goes through all of its amplitude values many times while one source wave arrives. Its effective amplitude is its average amplitude.

(2) Frequency$_{source}$ $\gg$ Frequency$_{encountered}$

The source center goes through all of its amplitude values many times while the encountered does once. Its effective amplitude is its average amplitude.

(3) Frequency$_{source}$ = Frequency$_{encountered}$

The interaction takes place over exactly one cycle and the effective amplitude is, again, the average.

In real matter, not the idealized model of one source and one encountered center, every *Spherical-Center-of-Oscillation* is constantly "bombarded" by various waves from a variety of directions at a variety of frequencies and phases due to the immense number of *Spherical-Centers-of-Oscillation* making up ordinary matter. The relative frequency and the phase of the wave and the encountered center have no effect on the large scale result from the interaction. Thus *Factor 2* is not a variable quantity but merely the average amplitude of the encountered center, which is designated U_c.

However, the absolute frequency of the encountered *Spherical-Center-of-Oscillation* is *Factor 3* in the formula for responsiveness. Just as the incoming wave repetition rate affects the amount of force that the wave can deliver to the encountered center, so the encountered center repetition rate affects that center's response to the wave. While the wave is encountering the center, each cycle of the encountered center's

28

oscillation is acted on by the wave. (This is most easily visualized if the frequency of the encountered center is much larger than that of the wave, but it applies in any case.)

Thus *Factor 3* is encountered center repetition rate. [For a center at rest the "rep rate" is the oscillation frequency but for a center in motion its velocity is a factor in the "rep rate" along with its oscillation frequency.

Then equation *3-5* becomes

(3-7)   responsiveness $\propto$ [cross-section]$\cdot$[amplitude]$\cdot$[rep rate]

$$= [K_{cs} \cdot \lambda_c^2] \cdot [U_c] \cdot [f_c]$$

$$= K_{cs} \cdot \lambda_c \cdot U_c \cdot c \qquad \text{[Using } c = f \cdot \lambda \text{]}$$

where: K_{cs} = a constant for the proportionality
          λ_c = the encountered center oscillation wavelength
          U_c = its amplitude, and
          f_c = its frequency.

PRECISE FORMULATION OF COULOMB'S LAW

The treatment here is of one single unit charge, $\pm U_c \cdot [1 - Cos(2\pi \cdot f \cdot t)]$, interacting with another such single unit charge, one simple basic *Spherical-Center-of-Oscillation* interacting with another.

The Encountered Center Charge Qe and Its Amplitude Uc

In the traditional formulation of Newton's Law equation *3-8*

(3-8)   Force = mass$\cdot$acceleration

and for the case that is now being considered, that in which the force results from the electrostatic interaction between two charges in accordance with Coulomb's Law, equation *3-9*,

(3-9)
$$\text{Force} = \frac{\text{Charge} \cdot \text{Charge}}{\text{Separation Distance}^2}$$

both of the charges enter into the relationship in the *Force* part, the *Mass* part of the relationship being like an inert characteristic of the substance.

In this *Centers-of-Oscillation* formulation equation *3-2*, repeated here

(3-2) $\begin{bmatrix} \text{Acceleration} \\ \text{Resulting} \end{bmatrix} = \begin{bmatrix} \text{Wave} \\ \text{Impulse} \end{bmatrix} \cdot \begin{bmatrix} \text{Responsiveness} \\ \text{of the Center} \end{bmatrix}$

or, more succinctly,

Acceleration = Wave × Responsiveness

the amplitude of the oscillation, U_c for the center, U_w for the wave, the role of which corresponds to that of traditional charge, Q, enters into the formulation differently from the traditional conception. The source *Spherical-Center-of-Oscillation*'s amplitude is a factor in the Wave and the encountered *Spherical-Center-of-Oscillation*'s amplitude is a factor in the Responsiveness.

Figure 3-3 on the following page compares the two.

Field and Wave, not Force and Wave, correspond. Each is the unrealized potential that becomes action via interaction with an encountered charge / center. Therefore the *[Charge ÷ Mass]* of the left half of Figure 3-3 is the same as the *Responsiveness* of the right half of the figure.

Traditional   *Centers - of - Oscillation*

$$\text{Acceleration} = \text{Force} \times \left[\frac{1}{\text{Mass}}\right] \qquad \text{Acceleration} = \left[\begin{array}{c}\text{Wave}\\\text{Im pulse}\end{array}\right] \times \text{Responseness}$$

$$= \left[\frac{Q \cdot Q}{d^2}\right] \times \left[\frac{1}{\text{Mass}}\right]$$

$$= \left[\frac{Q_s}{d^2}\right] \times \left[\frac{Q_e}{\text{Mass}}\right]$$

$$= \left[\begin{array}{c}\text{Electric}\\\text{Field at } d^2\end{array}\right] \times \left[\frac{Q_e}{\text{Mass}}\right] \qquad = \left[\begin{array}{c}\text{Wave}\\\text{Im pulse}\end{array}\right] \times \left[K_{cs} \cdot \lambda_c \cdot U_c \cdot c\right]$$

Figure 3-3

Therefore

(3-10)
$$\frac{Q_e}{m_e} = K_{cs} \cdot \lambda_c \cdot U_c \cdot c$$

from which

(3-11)
$$Q_e = \frac{h}{\lambda_c \cdot c}\left[K_{cs} \cdot \lambda_c \cdot U_c \cdot c\right] \qquad [\text{Using } m \cdot c^2 = h \cdot f]$$

$$= h \cdot K_{cs} \cdot U_c$$

which relates the charge of the encountered *Spherical-Center-of-Oscillation* to it's amplitude, and is a simple direct proportionality because h and K_{cs} are constants.

The Source Center Charge Q_s and Its Oscillation Amplitude U_c

If time could be stopped so that the waves from the source center were frozen in whatever position that they had in space, then the spherical waves as propagated by a *Spherical-Center-of-Oscillation* would appear as a series of nested shells, each of a successively greater radius, R, the radius being

(3-12) $\quad R_w = n \cdot \lambda_w$

where: $n = 1, 2, 3 \ldots$ for the successive shells
λ_w = the wavelength of the waves

and the thickness of each shell is the wavelength, λ_w. One such shell is depicted two-dimensionally in Figure 3-4, below.

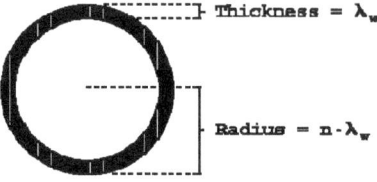

Figure 3-4

A cross-sectional view of this wave in space, that is a graph of its amplitude variation along a radius while traversing the thickness, is depicted in Figure 3-5, below, where it is clear that the area under the curve of amplitude variation is equal to $U_w \cdot \lambda_w$.

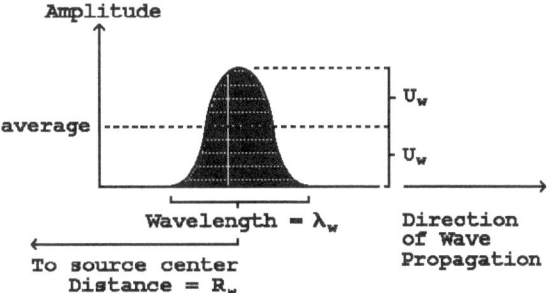

Figure 3-5

The potential impulse in one complete spherical shell, one wave cycle, is the shell cross-section, $U_w \cdot \lambda_w$, multiplied by the spherical surface area of the shell, $4\pi \cdot R_w$.

(3-13)   [a cycle of wave impulse] = $[U_w \cdot \lambda_w] \cdot [4\pi \cdot R_w]$

But, the wave amplitude, U_w, is the *Spherical-Center-of-Oscillation*'s amplitude, U_c, divided by the area of the wave's spherical shell at R_w and $\lambda_w = \lambda_c$ so that

(3-14)   [a cycle of wave impulse] = $U_c \cdot \lambda_c$

The *Wave* of Figure 3-3 is the equation *3-14* single *[a cycle of wave impulse]* multiplied by the repetition rate, the frequency, $f_w = f_c$, so that the wave, of Figure 3-3 is

(3-15)   Wave = $[U_c \cdot \lambda_c] \cdot f_c = U_c \cdot c = Q_s$,

which relates the field of the source *Spherical-Center-of-Oscillation* to that center's oscillation amplitude and, therefore, relates the charge of the source center to its amplitude.

Recognizing that every *Spherical-Center-of-Oscillation* is always in both source and encountered roles, then setting equation *3-11* equal to equation *3-14* the following is obtained.

(3-16) $Q_e = Q_s$

$\quad\quad\quad$ $h \cdot K_{cs} \cdot U_c = U_c \cdot c$

therefore

$\quad\quad\quad\quad$ $Q = U \cdot c$ $\quad$ and $\quad$ $K_{cs} = {}^c/_h$

Two Such Charges Interact Electrostatically As Follows

$\quad$ (1) The total potential force in the wave series as propagated by the source *Spherical-Center-of-Oscillation*s is (from equation *3-15*)

(3-15) $U_c \cdot c$

$\quad$ (2) The total wave series potential force per unit area of wave front at the encountered *Spherical-Center-of-Oscillation* is the quantity of step (1) divided by the spherical surface at the encountered center.

(3-17) $\dfrac{U_c \cdot c}{4\pi \cdot R^2}$

31

(3) The responsiveness of the encountered *Spherical-Center-of-Oscillation* is (equation $3\text{-}7$)

(3-7) $\text{Responsiveness} = K_{cs} \cdot \lambda_c \cdot U_c \cdot c$

(4) The resulting acceleration is, therefore (substituting steps (2) and (3), above, into equation $3\text{-}3$ per equation $3\text{-}6$)

(3-18)

$$\text{Acceleration} = \begin{bmatrix} \text{Wave Potential} \\ \text{Impulse per Unit} \\ \text{Area at the En-} \\ \text{countered Center} \end{bmatrix} \cdot \begin{bmatrix} \text{Responsiveness} \\ \text{of the} \\ \text{Encountered} \\ \text{Center} \end{bmatrix}$$

$$= \frac{U_c \cdot c}{4\pi \cdot R^2} \cdot K_{cs} \cdot \lambda_c \cdot U_c \cdot c$$

(5) The mass of the encountered *Spherical-Center-of-Oscillation* (from $m \cdot c^2 = h \cdot f$) is

(3-19)

$$m = \frac{h}{c \cdot \lambda_c}$$

(6) The force is, then (substituting steps (4) and (5), above into equation $3\text{-}2$)

(3-20) $\text{Force} = \text{Mass} \times \text{Acceleration}$

$$= \left[\frac{h}{c \cdot \lambda_c} \right] \cdot \left[\frac{U_c \cdot c}{4\pi \cdot R^2} \right] \cdot K_{cs} \cdot \lambda_c \cdot U_c \cdot c$$

$$= \frac{[U_c \cdot c] \cdot [h \cdot K_{cs} \cdot U_c \cdot c]}{4\pi \cdot R^2}$$

and substituting per 3-11 and 3-15 yields the result

(3-21)

$$\text{Force} = \frac{Q_s \cdot Q_e}{4\pi \cdot R^2}$$

which is Coulomb's law as it naturally occurs.

If a constant of proportionality, k, is introduced to accommodate choice of the units of charge, and the 4π is absorbed into that new constant, then the result (using q for charge since the added constant requires an accordingly different variable) is

(3-22)

$$\text{Force} = k \cdot \frac{q_s \cdot q_e}{R^2} \qquad k = {}^1\!/_{4\pi\varepsilon_0}$$

which is Coulomb's Law as originally formulated.

Here, Coulomb's Law is derived from the Nature of Matter, from the unavoidable requirements of the way the "Big Bang" started, not as a law inferred from empirical data as the Coulomb's Law of traditional 20th Century physics is.

$\longrightarrow$

$\longrightarrow$

SECTION 4

The Action of Matter: Motion and Relativity

THE PROBLEM

A *Spherical-Center-of-Oscillation* naturally sends a *Propagated Outward Flow* of *Medium* uniformly radially outward in all directions from itself at velocity c, the speed of light, as presented in Section 2. The speed of that flow is set by the μ_0 and ε_0 of the *Medium* to the exact value of c by virtue of their controlling the cyclical alternating exchange of the oscillation between the two forms in which it exists.

When the center is not in motion that presents no problem, but with the *Spherical-Center-of-Oscillation* moving in some direction the center's motion and its propagation are in conflict. In the direction of motion the velocity of the center, v, tends to add to the natural value of the speed, c, of propagation of the *Propagated Outward Flow* and in the opposite direction it tends to subtract. But, the speed of the flow is fixed; set at c by μ_0 and ε_0.

That conflict forces an adjustment of the oscillation of the *Spherical-Center-of-Oscillation* to modify its propagation speed of its *Propagated Outward Flow.*

THE *Spherical-Center-of-Oscillation at Constant Velocity*

The treatment is of the *Spherical-Center-of-Oscillation* at constant velocity because that is the more direct and simple case of motion, and at constant velocity one cannot detect absolute motion. That is, one can say that there is a relative difference of velocity between two systems at constant velocity in one of which the observer is located, but the observer cannot say which system is moving and which, if any, is at rest.

To describe the behavior of the center its propagation will be modeled resolved into three components: forward, rearward, and sideward relative to the direction of the center's velocity, as depicted in Figure 4-1. [In the figure the "up", "down", "left" and "right" are all "sideward".] These orthogonal components represent the propagated wave in all directions. The wave in any particular direction is the "resultant" of that directions' projection on the forward or rearward component (whichever is at a nearer angle) and on the sideward component. (The "resultant" is the hypotenuse of the right triangle having the projection components as its other two sides.)

35

When a center is at rest [absolute "rest" relative to its propagation] then propagation of waves is the same in all directions at speed $c = \lambda_r \cdot f_r$.

A Center-of-Oscillation at Rest

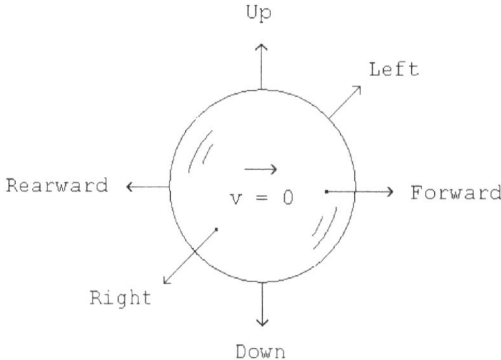

Figure 4-1

As described above under "The Problem" the speed of flow of centers' propagation is fixed at c by the μ_0 and ε_0 of the flowing *Medium*. The center moving at velocity v would find (in the forward direction) its freshly emitted propagation "thrown" forward at speed $[c + v]$ interfering with the flow just ahead of it at speed c and conflicting with the μ_0 and ε_0 of the *Medium*. It finds the propagated wave not moving out of the way at the needed $[c + v]$ in time for the next cycle as set by the at-rest frequency of the center. The result is an imperative to reduce the center frequency ["delay" the next cycle] by the factor $[1 - v/c]$. That "interfering" and "conflicting" tends to force on the center a change in its oscillation, a reduction by the factor $[1 - v/c]$. That is, with the center moving forward at v,

(4-1) Propagated Speed would become $c \cdot [1 - v/c] = (c - v)$

Flow speed = propagated speed + v = $(c - v) + v = c$

In the rearward direction the opposite is the case, an imperative to increase the center frequency by the factor $[1 + v/c]$. But, the *Spherical-Center-of-Oscillation* can only oscillate at one specific frequency at a time. It cannot both increase and decrease its oscillation frequency at the same time. It responds by adopting a compromise change in frequency, the geometric mean of the two conflicting factors as in equation 4-1.

The center's oscillation frequency decreases and its oscillation wavelength correspondingly increases, the product still being c.

(4-2)
$$f_v = f_r \cdot \left[1 - \frac{v^2}{c^2}\right]^{1/2}$$

[Center frequency decreases]

$$\lambda_v = \lambda_r \cdot \frac{1}{\left[1 - \dfrac{v^2}{c^2}\right]^{1/2}}$$

[Center wavelength increases]

$$\lambda_v \cdot f_v = \lambda_r \cdot f_r = c$$

[Wave velocity still at c]

While the center can oscillate at only one frequency, it can propagate at different wavelengths in different directions. To maintain propagated wave velocity at c in the direction of center motion the wave must be actually propagated forward by the center at

36

c-v relative to the center itself so that the wave velocity relative to at rest is the propagated velocity, *c*, plus the center velocity, *v*, that is *(c-v)+v = c.*

To propagate forward at *[c - v]* while maintaining the frequency at f_v requires that the wavelength change to a smaller value, λ_{fwd}. Likewise, rearward the wave must be actually propagated by the center at *[c + v]* relative to the center with a greater wavelength, λ_{rwd}. Those adjusted propagation wavelengths are as follows.

(4-3)

$$\lambda_{fwd} = \frac{c-v}{f_v} = \frac{c\cdot\left[1-\dfrac{v}{c}\right]}{f_r\cdot\left[1-\dfrac{v^2}{c^2}\right]^{\frac{1}{2}}} = \lambda_r \cdot \frac{\left[1-\dfrac{v}{c}\right]^{\frac{1}{2}}}{\left[1+\dfrac{v}{c}\right]^{\frac{1}{2}}} = \lambda_r \cdot \left[\frac{c-v}{c+v}\right]^{\frac{1}{2}}$$

$$f_{fwd} = \frac{c}{\lambda_{fwd}} = f_r \cdot \left[\frac{c+v}{c-v}\right]^{\frac{1}{2}}$$

$$\lambda_{rwd} = \lambda_r \cdot \left[\frac{c+v}{c-v}\right]^{\frac{1}{2}}$$

$$f_{rwd} = f_r \cdot \left[\frac{c-v}{c+v}\right]^{\frac{1}{2}}$$

The Wave as Propagated by the Center at Velocity v (relative to the center)

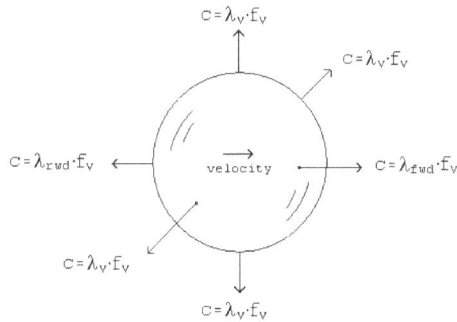

Figure 4-2

The Above Propagation (as Observed from At-Rest)

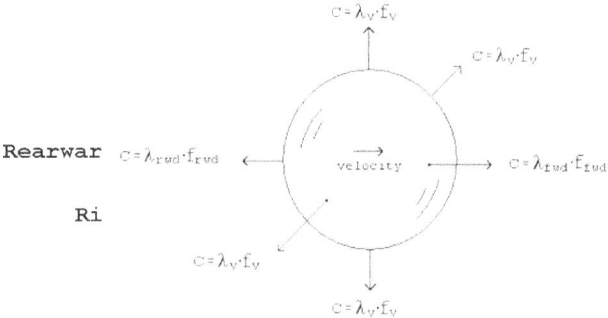

Figure 4-3

As the center "sees" it, per the above Figure 4-2, it is oscillating at f_v, with the forward and rearward wavelengths adjusted for the velocity so that the wave travels in each direction at speed **c**. As "at-rest" would "see" it, per Figure 4-3, below, the center appears to propagate different forward and rearward frequencies, f_{fwd} and f_{rwd}.

Thus the field of propagated waves is traveling at c in all directions as observed by the center that is in motion and doing the propagating and as observed from at-rest.

THE EFFECT OF VELOCITY ON MASS

With the oscillation frequency corresponding to the rest mass of the particle it represents per equations 2-6, the development so far of decreasing oscillation frequency, equation 4-2, demonstrates a decrease in rest mass due to the *Spherical-Center-of-Oscillation's* velocity. That is more properly referred to as a decrease in that part of the mass effect due to the overall frequency of oscillation of the center, to be referred to as "mass in rest form", m_r' in equation 4-3.

(4-4)
$$m'_r = m_r \cdot \frac{f_v}{f_r} = m_r \cdot \left[1 - \frac{v^2}{c^2}\right]^{\frac{1}{2}}$$

However, overall the total mass increases because the effects so far have reduced the cross-section target for interception of *Propagated Outward Flow*.

From the forward or the rearward point of view the center's cross-section is the area of the circle of radius λ_v, the sideward direction. Per equation 4-2.

(4-5)
$$\lambda_v = \lambda_r \cdot \frac{1}{\left[1 - \frac{v^2}{c^2}\right]^{\frac{1}{2}}}$$

Relative to the center's rest mass, m_r, the overall mass at velocity, m_v, is

(4-6)
$$m_v = m'_r \cdot \left[\frac{\lambda_v}{\lambda_r}\right]^2 = m_r \cdot \left[1 - \frac{v^2}{c^2}\right]^{\frac{1}{2}} \cdot \left[\frac{1}{\left[1 - \frac{v^2}{c^2}\right]^{\frac{1}{2}}}\right]^2$$

$$m_v = m_r \cdot \frac{1}{\left[1 - \frac{v^2}{c^2}\right]^{\frac{1}{2}}}$$

From the sideward point of view the cross-section is no longer a circle, however. In the forward direction the at-rest circle's radius has become λ_{fwd} instead of λ_v and in the rearward direction λ_{rwd} instead of λ_v.

(4-7)
$$\lambda_{fwd} = \frac{c - v}{f_v} = \frac{c \cdot \left[1 - \frac{v}{c}\right]}{f_v} = \frac{\left[1 - \frac{v}{c}\right]}{\lambda_v} \qquad \text{therefore} \qquad \frac{\lambda_{fwd}}{\lambda_v} = \left[1 - \frac{v}{c}\right]$$

$(4-8)$

$$\lambda_{rwd} = \frac{c+v}{f_v} = \frac{c \cdot \left[1 + \dfrac{v}{c}\right]}{f_v} = \frac{\left[1 + \dfrac{v}{c}\right]}{\lambda_v} \qquad \text{therefore} \qquad \frac{\lambda_{rwd}}{\lambda_v} = \left[1 + \frac{v}{c}\right]$$

The product of the change factors, equations $4-7$ and $4-8$, is $[1 - v^2/c^2]$, a reduction of cross-section, the same amount of increase in mass as equation $4-6$.

The other two factors in responsiveness beside cross-section, namely amplitude and repetition rate, are not further affected by velocity. Amplitude is unchanged and the affect of reduced oscillation frequency / repetition rate has been accounted for by the effect "mass in rest form" and equation $4-4$. Thus in all directions the effect of velocity is an increase in mass per equation $4-6$.

THE LORENTZ CONTRACTIONS, LENGTH AND TIME

Logic requires of the overall universe that in all frames of reference at constant velocities relative to each other [*i.e.* inertial frames]:

- The equations describing the laws of physics have the same form, and

- The universal constants appearing in those equations be the same,

This is called the Principle of Invariance, and means that the speed of light, c, a universal constant, is the same in all inertial frames, which appears to conflict with our instinctive assumption that the speed of light should vary with the speed of the light's source.

That logic combined with experiments showing that the speed of light actually is the same independent of whatever inertial frame, required the development of the Lorentz Transformations to account for the constancy of the speed of light. The transformations are coordinate transformations between two inertial frames. The Lorentz contractions are the related change in the fundamental quantities: mass, length, and time.

The Lorentz Contractions

The Lorentz Contractions are as follows.

$(4-9)$

$$L = L_r \cdot \left[1 - \frac{v^2}{c^2}\right]^{\frac{1}{2}} \qquad \begin{array}{l}\texttt{[Observed Length in the}\\ \texttt{\ Direction of motion shortens.]}\end{array}$$

$$f = f_r \cdot \left[1 - \frac{v^2}{c^2}\right]^{\frac{1}{2}} \qquad \texttt{[Observed frequency slows.]}$$

$$t = t_r \cdot \frac{1}{\left[1 - \dfrac{v^2}{c^2}\right]^{\frac{1}{2}}} \qquad \begin{array}{l}\texttt{[Observed time periods length,}\\ \texttt{\ Time passes more slowly.]}\end{array}$$

$$m = m_r \cdot \frac{1}{\left[1 - \dfrac{v^2}{c^2}\right]^{\frac{1}{2}}} \qquad \texttt{[Observed mass increases.]}$$

39

Time and frequency are reciprocals of each other and the above equation *4-2* decrease in center frequency with velocity validates the f and t Lorentz Transforms. [The increasing λ_r to λ_v of that equation is compensating for the frequency decrease to keep the sideward propagation speed at c. Sideward is not the direction of v so the Lorentz Contraction does not apply to that λ.]

The equation *4-6* overall increase in center mass with velocity validates the mass, m, Lorentz Transform. Remaining to be validated is the length, L contraction. The λ_{fwd} and λ_{rwd} contraction equations *4-7* and *4-8* are a center length contraction in the velocity direction, a Lorentz Contraction.

On the macroscopic scale it is necessary to investigate two centers and the distance between them in order to develop a velocity-caused contraction of length in matter. In bulk matter composed of multiple particles, atoms and their components, the spacing of the atoms depends on the balance of the various electrostatic forces acting as a result of the centers-of-oscillation, protons and electrons, of which the matter atoms are composed.

Considering just two *Spherical-Centers-of-Oscillation* at rest in a fixed position relative to each other, the effect of their moving jointly at velocity v in the direction of the line joining them should be a Lorentz Contraction to closer spacing of the two centers by the Lorentz Contraction factor.

The position of each of the two centers is the balance of all of the forces acting on the centers, an equilibrium position. If the velocity is to change the distance between the two centers then the force acting between the two centers must change so that a new closer equilibrium spacing exists and determines the new distance between the two centers. For the centers to need to be closer in order to re-establish equilibrium the effective charge of each of the centers must be decreased by the velocity.

In other words, for the Coulomb force between the two centers

(3-21)
$$F = \frac{Q_1 \cdot Q_2}{4\pi \cdot R^2}$$

to be unchanged even though R is reduced by the Lorentz Contraction by the factor

(4-10)
$$\frac{R_{vel}}{R_{rst}} = \left[1 - \frac{v^2}{c^2}\right]^{\frac{1}{2}}$$

so that R^2 is changed by the factor

$$\frac{R^2_{vel}}{R^2_{rst}} = \left[1 - \frac{v^2}{c^2}\right]$$

then $Q_1 \cdot Q_2$ must be so reduced by the same factor as is R^2.

But, that is exactly the case. It has already been shown by equation *4-3* that the forward wave propagation speed is reduced by the factor *[1-^{v}/c]* to $c' = c-v$ and that the rearward wave propagation speed is analogously changed by the factor *[1+^{v}/c]* to $c'' = c+v$.

40

From equation *2-16* $Q = U \cdot c$ so that the charge, Q, of the trailing center "looking" forward is reduced by the reduction of its c to $c' = c - v$, a factor of $[1 - V/c]$.

Similarly the charge, Q, of the leading center "looking" backward is increased by the increase of its c to $c'' = c + v$, a factor of $[1 + V/c]$.

Therefore, $Q_1 \cdot Q_2$ is reduced by the product of the two factors which is $[1 - v^2/c^2]$, which matches the Lorentz Contraction of R^2 and therefore of R and validates the length, L, Lorentz Contraction.

The Velocity-Caused Impulse Increment

There is, however, another component to the interaction. While, in the forward direction, the source *Spherical-Center-of-Oscillation* propagates the wave at $c' = c - v$, the wave actually travels at velocity c because the center itself is traveling forward at v yielding the overall wave velocity as $c' + v = (c - v) + v = c$. The forward wave, the force it can deliver reduced by its propagation at $c - v$, is thus also "thrown forward" by the center's velocity. This adds another component of force, of potential impulse per wave times the wave repetition rate, the force that the wave can deliver to an encountered center.

In fact, without the wave having that additional component of force, and the consequent reaction back on the center in that same additional amount, the center would not experience equal reaction back on it in all directions from its propagated wave. The magnitude of this "force component" due to the center's velocity is $[V/c] \cdot F_r$, where F_r is the force that the wave would deliver if at rest and which it does always deliver to the sides: up, down, right and left.

The situation is analogous for the rearward wave otherwise the reaction back on the center by the rearward propagated wave would be $[V/c] \cdot F_r$ greater than the rest case. Without these "force components" the center would be self-accelerated in the forward direction by a force of $2 \cdot [V/c] \cdot F_r$ (the forward and rearward effects combined), clearly not the case.

Returning to the case of two *Spherical-Centers-of-Oscillation* traveling in the direction of an imaginary line joining them, when the forward wave of the trailing center encounters the rear of the leading center the $+[V/c] \cdot F_r$ positive "force component" of the forward wave and the $-[V/c] \cdot F_r$ negative "force component" of the rear of the encountered leading center cancel out leaving the net action due to the encounter as presented above before considering the "force component due to center velocity or momentum" aspect.

The situation is the same with the rearward propagated wave of the leading center encountering the front of the trailing center. The net effect on the interaction is null, but the phenomena are still there.

Particle Kinetics

Kinetic energy, KE, is defined as the work done by the force, f, acting on the particle or object of mass, m, over the distance that the force acts, s. This quantity is calculated by integrating the action over differential distances. It was done using the Lorentz Contraction for mass originally by Einstein as follows

(4-10)
$$KE = \int_0^s f \cdot ds \qquad \text{[Per above definition]}$$

$$= \int_0^s \frac{d(m \cdot v)}{dt} \cdot ds \qquad \begin{array}{l}\text{[Newton's 2}^{nd}\text{ law,]} \\ \text{[f = m} \cdot \text{a = m} \cdot dv/dt\text{]}\end{array}$$

$$= \int_0^{(m \cdot v)} \frac{ds}{dt} \cdot d(m \cdot v) \qquad \text{[Rearrangement of form]}$$

$$= \int_0^{(m \cdot v)} v \cdot d(m \cdot v) \qquad \text{[v = }ds/dt\text{]}$$

But, now the mass, *m*, increases with velocity per equation *4-9*, Therefore:

$$KE = \int_0^v v \cdot d \left[\frac{m_r \cdot v}{\left[1 - \dfrac{v^2}{c^2} \right]^{\frac{1}{2}}} \right] \qquad \begin{array}{l}\text{[m is m}_r\text{ Lorentz} \\ \text{contracted by v.} \\ \text{m}_r\text{ is rest mass]}\end{array}$$

$$= \frac{m_r \cdot v^2}{\left[1 - \dfrac{v^2}{c^2} \right]^{\frac{1}{2}}} - m_r \cdot \int_0^v \frac{v \cdot dv}{\left[1 - \dfrac{v^2}{c^2} \right]^{\frac{1}{2}}} \qquad \begin{array}{l}\text{[Integration} \\ \qquad \text{by parts]}\end{array}$$

(4-11)
$$KE = \frac{m_r \cdot v^2}{\left[1 - \dfrac{v^2}{c^2} \right]^{\frac{1}{2}}} + m_r \cdot c^2 \cdot \left[1 - \frac{v^2}{c^2} \right]^{\frac{1}{2}} - m_r \cdot c^2 \qquad \begin{array}{l}\text{[Integration} \\ \text{of 2nd term]}\end{array}$$

$$= \frac{m_r \cdot v^2 + m_r \cdot c^2 \cdot \left[1 - \dfrac{v^2}{c^2} \right]}{\left[1 - \dfrac{v^2}{c^2} \right]^{\frac{1}{2}}} - m_r \cdot c^2 \qquad \begin{array}{l}\text{[Place 2}^{nd}\text{ term} \\ \text{over 1}^{st}\text{ term} \\ \text{denominator]}\end{array}$$

$$= \frac{m_r \cdot v^2 + m_r \cdot c^2 - m_r \cdot v^2}{\left[1 - \dfrac{v^2}{c^2} \right]^{\frac{1}{2}}} - m_r \cdot c^2 \qquad \begin{array}{l}\text{[Expand term} \\ \text{with brackets]}\end{array}$$

42

$$KE = \frac{m_r \cdot c^2}{\left[1 - \frac{v^2}{c^2}\right]^{\frac{1}{2}}} - m_r \cdot c^2 \qquad \text{[Simplify]}$$

$$= m_v \cdot c^2 - m_r \cdot c^2$$

[m_v is total mass at $v > 0$

m_r is total mass at $v = 0$

$m_v = m_r$ Lorentz transformed]

This result states that:

{Kinetic Energy} = {Total Energy} - {Rest Energy}

or

{Total Energy} = {Kinetic Energy} + {Rest Energy}

The appearance in this result that the energies are the product of the masses times c^2, the speed of light squared, was the origination of that concept, the famous Einstein's $E = m \cdot c^2$. The concept falls out naturally from applying the Lorentz transforms to the classical definition of kinetic energy. It is somewhat surprising that Einstein was the first to do that inasmuch as it was Lorentz who developed the Lorentz transforms and the Lorentz contractions.

Alternative Treatment of the Same Derivation

If in the above original derivation one proceeds differently from the first line of equation *4-11* on, as below, a slightly different result is obtained.

(4-11)
$$KE = \frac{m_r \cdot v^2}{\left[1 - \frac{v^2}{c^2}\right]^{\frac{1}{2}}} + m_r \cdot c^2 \cdot \left[1 - \frac{v^2}{c^2}\right]^{\frac{1}{2}} - m_r \cdot c^2 \quad \text{[Repeat \textit{(4-11)}}$$
$$\text{to start here]}$$

(4-12)
$$KE + m_r \cdot c^2 = \frac{m_r \cdot v^2}{\left[1 - \frac{v^2}{c^2}\right]^{\frac{1}{2}}} + m_r \cdot c^2 \cdot \left[1 - \frac{v^2}{c^2}\right]^{\frac{1}{2}} \quad \text{[Move the}$$
$$\text{right most}$$
$$\text{"- } m_r \cdot c^2 \text{"]}$$

Considering and evaluating the three terms of equation *(4-12)*:

$KE + m_r \cdot c^2$ = Kinetic plus rest energies

= Total Energy

= $m_v \cdot c^2$

$$\frac{m_r \cdot v^2}{\left[1 - \frac{v^2}{c^2}\right]^{\frac{1}{2}}} = \text{A relativistically increased}$$
energy of motion.

= $m_v \cdot v^2$

43

$$m_r \cdot c^2 \cdot \left[1 - \frac{v^2}{c^2}\right]^{\frac{1}{2}} = \text{A relativistically reduced}$$
$$\text{rest energy.}$$
$$= m_v \cdot c^2 - m_v \cdot v^2$$

the result is that equation *4-12* is equivalent to

$$(4\text{-}13) \quad \begin{bmatrix} \text{Total} \\ \text{Energy} \end{bmatrix} = \begin{bmatrix} \text{Energy in} \\ \text{Kinetic Form} \end{bmatrix} + \begin{bmatrix} \text{Energy in} \\ \text{Rest Form} \end{bmatrix}$$

$$m_v \cdot c^2 \quad = \quad m_v \cdot v^2 \quad + \quad m_v \cdot (c^2 - v^2)$$

and (dividing the above energy equation by c^2 to obtain an equation in mass)

$$(4\text{-}14) \quad \begin{bmatrix} \text{Total} \\ \text{Mass} \end{bmatrix} = \begin{bmatrix} \text{Mass in} \\ \text{Kinetic Form} \end{bmatrix} + \begin{bmatrix} \text{Mass in} \\ \text{Rest Form} \end{bmatrix}$$

$$m_v \quad = \quad m_v \cdot v^2 / c^2 \quad + \quad m_v \cdot (1 - v^2 / c^2)$$

The m'_r "mass in rest form", of equation *4-4*

$$m'_r = m_r \cdot \left[1 - \frac{v^2}{c^2}\right]^{\frac{1}{2}}$$

equals the m_v of equation *4-6*, below $\qquad$ multiplied by the $(1 - v^2/c^2)$ as in the bottom of equation *4-14*

$$m_v = m_r \cdot \frac{1}{\left[1 - \frac{v^2}{c^2}\right]^{\frac{1}{2}}} \qquad\qquad \left[1 - \frac{v^2}{c^2}\right]$$

Why is the formulation for classical *Kinetic Energy* $KE = \frac{1}{2} \cdot m \cdot v^2$ but *Energy in Kinetic Form* is simply $m \cdot v^2$ without the $\frac{1}{2}$? When dealing with quite small velocities (v very small relative to c) the excursion of total energy above rest energy and the excursion of energy in rest form below rest energy are both essentially linear. In that case the portion above the rest case is essentially half of the total excursion above and below the rest case. The classical kinetic energy is then half, $\frac{1}{2} \cdot m \cdot v^2$, the total energy in kinetic form, $m \cdot v^2$, for $[v/c]$ quite small.

THE CENTER OF OSCILLATION "AT REST" AND "IN MOTION"

In motion at a constant velocity, v, the *Spherical-Center-of-Oscillation* experiences the asymmetrical distortions of equation *4-3* and figures 4-2 and 4-3. The distortions indicate the motion and the motion enhanced energy of the center. At rest, in the absence of motion the center is spherically symmetrical.

Thus the rest mass and rest energy correspond to the spherically symmetrical portion of the center's oscillation [the only portion if v = 0] and they are "mass in rest form" and "energy in rest form". The overall distorted portion corresponds to the total "mass in kinetic form" and "energy in kinetic form" of the center. Of course the difference of the two is the mass and energy in kinetic form.

This brings up the point that, contrary to Einstein, there is an absolute frame of reference, an "at rest" frame. When the *Spherical-Center-of-Oscillation*'s oscillation is perfectly spherically symmetric then the center's velocity is zero and it is completely at rest. That is the universe's absolute frame to which all motion and all other frames are relative.

That is why the the Principle of Invariance is valid. That is why it is required of the overall universe that in all frames of reference at constant velocity relative to each other [*i.e.* inertial frames]:

- The equations describing the laws of physics have the same form, and

- The universal constants appearing in those equations are the same,

- The speed of light, c, a universal constant is the same everywhere.

They are all part of the same one overall absolute frame of reference, the rest frame. The rest frame is not special in that its laws and constants are not different. It is special in that all other frames are relative to it. It is simply the actual frame of the "Big Bang".

The Action of Matter: The Magnetic Effect
Ampere's Law

When it is at rest, the interaction of a *Spherical-Center-of-Oscillation* with its environment, other centers, is the electrostatic Coulomb effect. But, when the center is in motion, there is an additional effect, the magnetic effect.

STATIC MAGNETIC BEHAVIOR

Static magnetic behavior is the interactive effect of one center on another, due to both at constant velocity. A center so in motion exerts on another center a second force due to the motion. The direction of this interactive force is summarized in Figure 5-1 on the following page: a pair of electric currents (flows of electric charges, flows of *Spherical-Centers-of-Oscillation*) in various orientations relative to each other.

There are five basic cases of relative orientation of the two currents interacting. Any other situation can be resolved into some combination of them. In each case the analysis is of the effect of current #1 on current #2. Of course, exactly analogous reasoning would treat the effect of current #2 on #1.

Both effects occur simultaneously just as in the earlier discussion of Coulomb's Law each center is in both source and encountered roles simultaneously even though the action is described in terms of only one of the roles at a time. I is the commonly used symbol for current. F_M is the magnetic force.

Electric current being a flow of electric charges, one can speak of current or of some-quantity-of-charges-with-some-velocity. The following discussion must use both terminologies in order to relate the one to the other. A positive current in a given direction corresponds to positive charges flowing in that direction and equally corresponds to negative charges flowing in the opposite direction. Each charge is a *Spherical-Center-of-Oscillation.*

The magnitude of the static magnetic force, F_M, is given by Ampere's Law, equation *5-1* for Case #1 or #2.

(5-1)
$$F_M = \frac{\mu}{2\pi} \cdot \frac{I_1 \cdot I_2}{R} \cdot L$$

where: L = the length of each of the two parallel current paths
over which the force acts,

R = the distance between the two parallel paths,

μ = the permeability, a magnetic parameter of the space
between the two current paths (as before in the
discussion of the velocity of light).

47

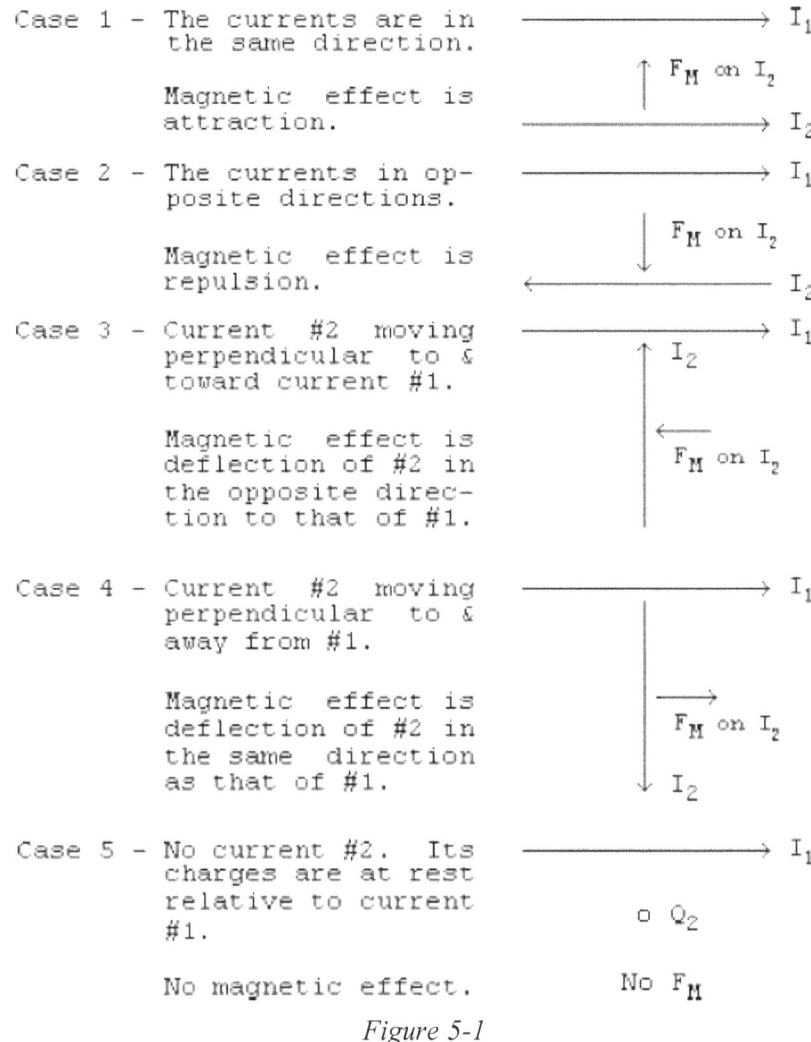

Figure 5-1

The magnetic effects are in addition to the electrostatic (Coulomb) effects of the same charges. That is, the charges, the *Spherical-Centers-of-Oscillation* whose motion is the current that produces the magnetic effects, have their natural (Coulomb) effect on each other when in motion as well as when at rest.

However, their motion changes the amount and direction of that effect and that change is the magnetic effect. To evaluate the magnetic force, then, it is necessary to examine the changes caused in the electrostatic force by the motion of those charges. If F_T is defined as the total interaction force, the combined effect of the electrostatic force, F_E, as it would be for those charges at rest and the magnetic force, F_M, due to their motion is, then

$(5-2)$ $\mathbf{F_T = F_E + F_M}$

where: F_T = the total interaction force between the charges,

F_E = the Coulomb force when the charges are at rest,

F_M = the magnetic force,

48

where the bold type indicates that the quantities have both magnitude and direction (are "vector" quantities) and both the magnitude and direction must be taken into account.

The analysis must therefore be: first an evaluation of the interactive force with the current's charges at rest, F_E ; then an evaluation of F_T , the interactive force in the same configuration but with the charges in motion at velocity, v; and, finally, the comparison of those two results to obtain the magnetic effect, F_M.

The natural geometry of the situations to be analyzed, the cases of Figure 5-1, makes the overall problem quite complicated. There is a great variety of configurations: the spherical form of each charge, the cylindrical symmetry of the currents and such currents in some cases perpendicular to each other, resulting magnetic forces in third directions, etc. As a result there is almost no way to obtain simple mathematical descriptions and analyses of what is actually happening. The real physical processes are direct and simple, but whether the mathematics is performed in rectangular, spherical or cylindrical coordinates some of the aspects of the problem will not be conveniently accommodated so that the mathematical expressions tend to become difficult.

Dividing equation $5-1$ by the path length, L , which is the length of current path I_1, the magnetic force of I_1 (the effect of its entire path length) on a unit length of I_2 (a minute length increment) is as in equation $5-3$, below.

$(5-3)$
$$F_M = \frac{\mu}{2\pi} \cdot \frac{I_1 \cdot I_2}{R} \qquad \text{[per unit length.]}$$

For equations $5-1$ and $5-3$ to be valid the actual path length, L, must be much greater than the minute length increment so as to prevent "end" effects. Theoretically the two current paths are infinitely long and a short section in the middle is being considered. That is, equation $5-3$ expresses the magnetic effect of all of the electric current of I_1 over its entire path length, L (where because the path of I_1 is so long the effect of the distant parts is negligible), acting on a unit length section of I_2.

Each of the currents, I_1 and I_2, is a stream of charges moving at velocity, v, as shown in Figure 5-2, below.

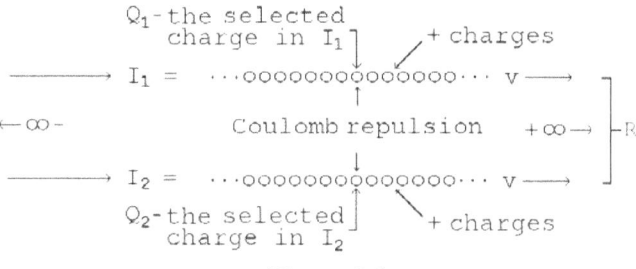

Figure 5-2

If Q_1 is the increment of charge in a unit length of the stream of charges that is I_1 and Q_2 is that for I_2 , then, per Coulomb's law, the electrostatic force that acts on charge increment, Q_2 , in I_2 due to the directly opposite charge increment, Q_1 , in I_1 is

$(5-4)$
$$F = \frac{Q_1 \cdot Q_2}{4\pi \cdot \varepsilon \cdot R^2} \qquad \text{[Coulomb's Law.]}$$

Of course, Q_2 is similarly affected by all of the other charges in I_1 , not just Q_1 .

Since length increments and force per unit length are being treated, the increments of charge Q_1 and Q_2 must be replaced with charge-per-unit-length, ρ_1 and ρ_2, so that, when multiplied by length increments charge amount is obtained.

The amount of charge, Q , located in length increment, dx , is

(5-5) $Q = \rho \cdot dx$

To analyze the total electrostatic effect of all of the charges of which I_1 is composed on a single charge increment in I_2 the analysis is as in Figure 5-3. In the figure R and $R(x)$ are radial distances between charge increments $Q = \rho \cdot dx$, that is R and $R(x)$ are the radial charge separation distance that appears in the denominator of Coulomb's law. The $dF(x)$ is the incremental Coulomb effect of a charge increment in I_1 on the selected charge increment in I_2 . F_r is the peak value that $dF(x)$ attains (when its $[R(x)]^2$ in the Coulomb's law denominator is a minimum, R^2).

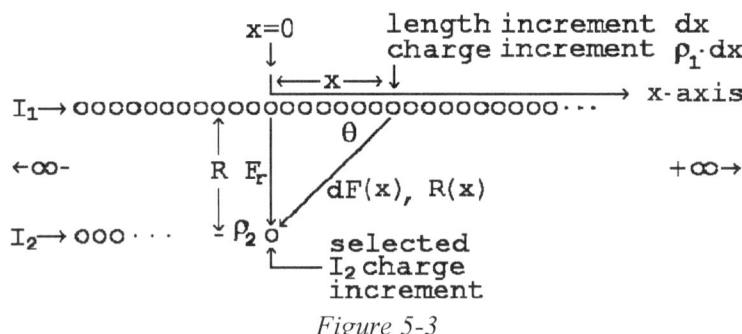

Figure 5-3

From Coulomb's law, the magnitude of $dF(x)$, the increment of the force, $F(x)$, exerted on the charge increment ρ_2 by $(\rho_1 \cdot dx)$, the amount of charge of I_1 that is located in length increment, dx , is

(5-6) $$dF(x) = \frac{(\rho_1 \cdot dx) \cdot \rho_2}{4\pi \cdot \varepsilon \cdot [R(x)]^2}$$

When $x=0$ so that $R(x)=R$ the rate of $dF(x)$ per dx is the pure "sideward" value, which equals the rest value and will be defined as F_r ,

(5-7) $$\frac{dF(x)}{dx} = \frac{\rho_1 \cdot \rho_2}{4\pi \cdot \varepsilon \cdot R^2} \equiv F_r$$

so that equation 5-6 then becomes

(5-8) $$dF(x) = F_r \cdot dx \cdot \frac{R^2}{[R(x)]^2}$$

From the right triangle $[\rho_2 - dx - x=0]$ in Figure 5-3 the sides of which triangle are R, x, and the hypotenuse, $R(x)$,

(5-9) $R(x) = \sqrt{x^2 + R^2}$ [law of Pythagoras]

so that equation 5-8 becomes

(5-10) $$dF(x) = F_r \cdot \frac{R^2}{x^2 + R^2} \cdot dx$$

This force magnitude is directed diagonally to the lower left in Figure 5-3. That is, the charges in I_1 and I_2 are of the same sign and repel each other. The charge increment of I_1 at dx repels ρ_2 as shown in the figure. Depending on which charge increment of I_1 is considered, the angle at which the force increment, $dF(x)$, acts varies.

Consequently, the analysis must be broken down into two orthogonal components. In terms of Figure 5-3 those components will be "horizontal", that is parallel to the figure's x-axis, and "vertical", at right angles to "horizontal". A quantity annotated with a horizontal arrow above it will mean that the "horizontal" component of the overall vector quantity is being treated. A quantity annotated with a vertical arrow to its left will mean that the "vertical" component is being treated.

The magnitudes of the components, $\overrightarrow{dF}(x)$ and $\uparrow dF(x)$ relate to the overall vector quantity, $\mathbf{dF(x)}$, as

(5-11) $\overrightarrow{dF}(x) = \mathbf{dF(x)} \cdot \dfrac{x}{R(x)} = \mathbf{dF(x)} \cdot \dfrac{x}{[x^2 + R^2]^{\frac{1}{2}}}$

$\uparrow dF(x) = \mathbf{dF(x)} \cdot \dfrac{R}{R(x)} = \mathbf{dF(x)} \cdot \dfrac{R}{[x^2 + R^2]^{\frac{1}{2}}}$

CASES 1,2 AND 5

The Static Force (The Charges At Rest)

From $x=-\infty$ to $x=0$ the horizontal components are all directed to the right. From $x=0$ to $x=+\infty$ they are all directed to the left. Summed up from $x=-\infty$ to $x=+\infty$ the horizontal components cancel out.

(5-12) $\overrightarrow{F}_E = 0$

If the vertical components, $\uparrow dF(x)$, are summed over that range the result will be the total electrostatic force of the charges in I_1 on a single charge increment in I_2, the force being sought.

Substituting $dF(x)$, equation 5-10 for $\mathbf{dF(x)}$ in the expression for $\uparrow dF(x)$, equation 5-11, yields the increments to be summed over the range $x=0$ to $+\infty$.

(5-13) $\uparrow dF(x) = F_r \cdot \dfrac{R^3}{[x^2 + R^2]^{1\frac{1}{2}}} \cdot dx$

It will become necessary to treat the region to the left, the [$x \le 0$] region, separately from the region to the right, the [$x \ge 0$] region. The analysis is restated as two problems, one for the range $x=-\infty$ to 0 and one for the range $x=0$ to $+\infty$.

See Appendix D, *Integration Details for Magnetic Effect Calculations, Part 1.*

(5-14) $\uparrow F_E = \displaystyle\int_{-\infty}^{0} \uparrow dF(x) + \int_{0}^{+\infty} \uparrow dF(x)$

$= \displaystyle\int_{-\infty}^{0} F_r \cdot \dfrac{R^3}{[x^2 + R^2]^{1\frac{1}{2}}} \cdot dx + \int_{0}^{+\infty} F_r \cdot \dfrac{R^3}{[x^2 + R^2]^{1\frac{1}{2}}} \cdot dx$

$= \pm R \cdot F_r + \pm R \cdot F_r = 2 \cdot R \cdot F_r$

By substituting in the value of F_r from equation 5-7, one obtains

(5-15) $\uparrow F_E = 2 \cdot R \cdot F_r = 2 \cdot R \cdot \dfrac{\rho_1 \cdot \rho_2}{4\pi \cdot \varepsilon \cdot R^2} = \dfrac{\rho_1 \cdot \rho_2}{2\pi \cdot \varepsilon \cdot R}$

The ratio of the magnitudes, F_M/F_E, is the F_M of equation 5-3 divided by the F_E of equation *5-13*.

(5-16)
$$\frac{F_M}{F_E} = F_M \cdot \frac{1}{F_E} = \left[\frac{\mu}{2\pi} \cdot \frac{I_1 \cdot I_2}{R}\right] \cdot \left[\frac{2\pi \cdot \varepsilon \cdot R}{\rho_1 \cdot \rho_2}\right] = \mu \cdot \varepsilon \left[\frac{I_1 \cdot I_2}{\rho_1 \cdot \rho_2}\right]$$

Since current is charge flow per unit time, then

(5-17) $I_1 = \rho_1 \cdot v_1$ and $I_2 = \rho_2 \cdot v_2$

where v_1 and v_2 are the velocities of the charges in I_1 and I_2. Using these and, letting $v_1 = v_2 = v$ for simplicity then

(5-18)
$$\frac{F_M}{F_E} = \mu \cdot \varepsilon \left[\frac{[\rho_1 \cdot v_1] \cdot [\rho_2 \cdot v_2]}{\rho_1 \cdot \rho_2}\right] = \mu \cdot \varepsilon \cdot v^2$$

is obtained. Finally recognizing that $c^2 = \frac{1}{\mu \cdot \varepsilon}$ and canceling the identical $\rho's$ the following is obtained.

(5-19)
$$\frac{F_M}{F_E} = \frac{v^2}{c^2} \qquad \text{or} \qquad F_M = \frac{v^2}{c^2} \cdot F_E$$

Although this analysis was performed only for Case 1, it is valid for all 5 cases. The magnitude of the magnetic force is the same (for analogous values of the currents, etc.) in all of Cases 1 - 4, and is zero, of course, for Case 5. Likewise, the simplifying assumption that the velocity of the charges in each of the current flows is the same does not change the general validity.

The analysis so far, while developing a somewhat new result and taking a somewhat new point of view, is nevertheless entirely a result of and performed in terms of traditional 20th Century physics. The relationship, equation *5-18*, expresses the magnitude of the change to the electrostatic effect that produces the magnetic effect.

The Magnetic Force (The Charges In Motion)

Now it is necessary to investigate how this comes about from the actions of *Spherical-Centers-of-Oscillation*. In the above analysis, while the velocity of the charges was indicated in the figure, the velocity was taken to be zero for F_E and the magnetic effect, F_M, was obtained from equation *5-2*, the traditional Ampere's Law. The analysis that produced F_E must now be performed again but with treating the charges as *Spherical-Centers-of-Oscillation* and modifying $F(x)$ (and, of course, $dF(x)$) as appropriate to the behavior of centers in motion at velocity v.

The magnetic force, alone, cannot be independently calculated. Rather, it can only be found by calculating the total interactive force with the charges in motion, F_T, and subtracting from that the portion that occurs when the charges are not in motion, the F_E just obtained. That is, from equation *5-2*,

(5-20) $\mathbf{F_T} - \mathbf{F_E} = \mathbf{F_M}$

Since velocity is now also a variable the symbols $F(x)$ and $dF(x)$ will be replaced with $F(v,x)$ and $dF(v,x)$. Subtracting F_E of equation *5-13* from this new F_T, the F_M portion can be obtained (taking account of direction, i.e. a vector subtraction by components). The magnitude of that F_M should be the same as obtained from Ampere's Law and that is most easily verified by taking the magnitude ratio F_M/F_E, which should be the same as equation *5-18*, which was obtained using the methods of traditional 20th Century physics.

As developed in the preceding Section 4, *The Action of Matter: Motion and Relativity*, the motion of a center at constant velocity results in changes in the propagated wave and in the center's own oscillation. Of interest here is the effective value of the charge. It is the variation in the effective value of the charge, Q ($\rho \cdot dx$), entering into the calculation of the net force effect per equation *5-6*, that produces the change.

Now, however, unlike the development in and following equation *5-6*, Q is not a constant so that F_r is not constant and this variation due to velocity must be included in the expression for $F(x)$ (and $dF(x)$) as used in equation *5-14*, above, namely the new quantity $F(v,x)$ (and $dF(v,x)$).

In the prior section it was shown that the forward propagated wave is reduced by the factor $[1-v/c]$ because of the forward propagation at $c' = c-v$ and that the rearward propagated wave is analogously changed by the factor $[1+v/c]$. It was also shown that the "throwing forward" of the forward wave by the center's velocity and the "negative" of that for the rearward wave changed the net force effect of the wave by a "force component" equal to $[v/c] \cdot F_r$, positive in the forward direction and negative in the rearward direction (where F_r is the force delivered at the same distance but from the center at rest or to its side).

These changes are summarized in Figures 5-4 and 5-5 which depict the wave propagated by the source center, Q_1 and the encountered center, Q_2 in the present analysis. The "force component" due to the center's velocity is F_{fc}. Since the force effect of the wave and the center is directly proportional to the charge, the effects developed in Section *4 The Action of Matter: Motion and Relativity* can be treated as changes to the effective force.

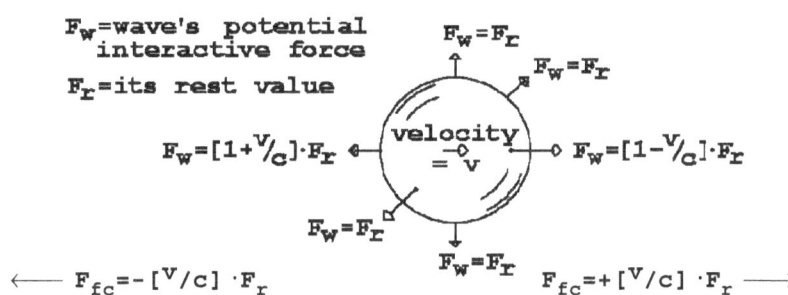

Figure 5-4
The Propagated Wave from the Source Center at Velocity v

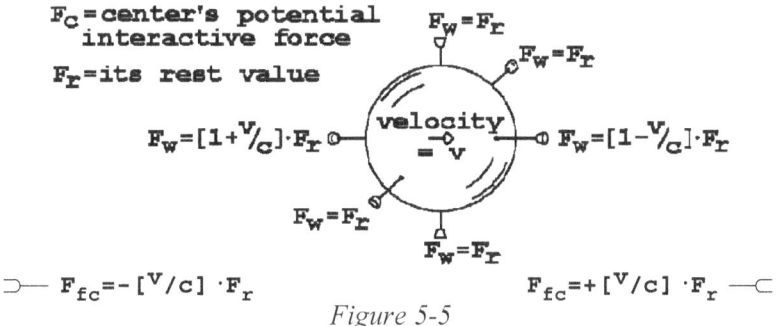

Figure 5-5
The Encountered Center's Oscillation at Velocity v

53

In motion as the currents I_1 and I_2 the centers exhibit cylindrical symmetry around their direction of motion so that a two dimensional analysis will suffice for the following determination of the actual force effect in any particular direction.

Figure 5-6, below, illustrates the difference between conditions at rest and at velocity. At rest the centers' oscillation and waves have the same force effect in all directions. At velocity the force effect depends upon the angle of view relative to the direction of the velocity, angle θ in the figure.

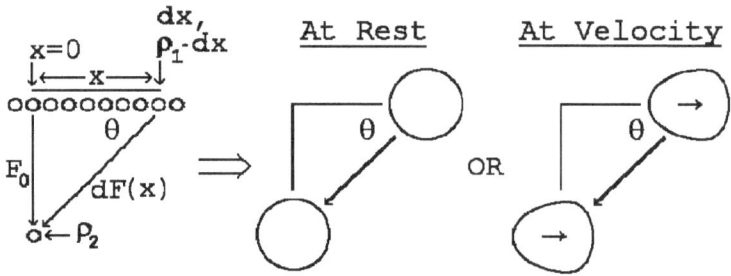

Figure 5-6
The Centers of Figure 5-2 Enlarged

For F_{fc}, the "force component", the analysis of the effect of the "angle of view", θ, is simple. As shown in Figure 5-7, below, its magnitude in any direction is equal to the cosine of the angle between that direction and the pure forward direction times $[v/c] \cdot F_r$. That relationship applies to both the F_{fc} of the wave and of the center (the point of view of Figure 5-4 and Figure 5-5, above).

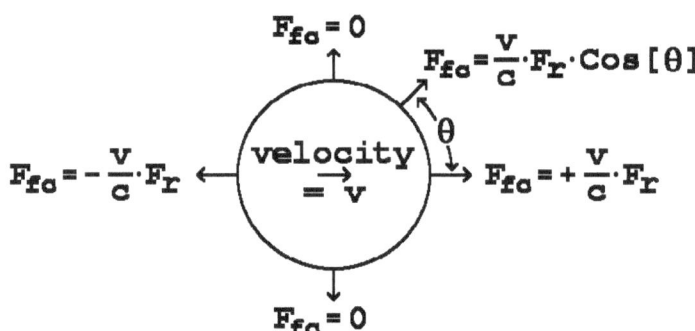

Figure 5-7
The "Force component" Resultant When the Center is at Constant Velocity v

The treatment for the variation of the effect with the angle of view, θ, being essentially the same for both the wave and the center, is also true for the forward wave propagation at $c'=c-v$ producing a reduction of the forward force effect by $[1-v/c]$.

For the wave, for which the four two-dimensional components (using the cylindrical symmetry of the situation) are per Figure 5-4, the situation is not so simple as for the F_{fc}. If the velocity were zero then the wave resultant would be F_r in all directions and the model of it would be a circle as in Figure 5-8, below. In any non-orthogonal direction the force is obtained from the law of Pythagoras; the force is the hypotenuse and the other two sides are its projection on the $x-$ and $y-axes$.

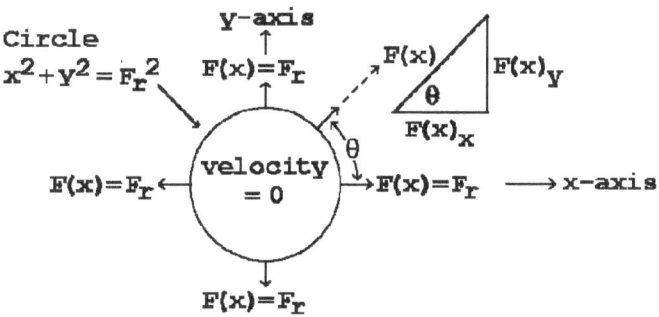

Figure 5-8
The Wave Resultant When the Center is at Rest

In the above figure, the center being at rest and its force effect being the same in all directions, *F(x)* is always equal to F_r regardless of θ. However, when the center is in motion the situation is analogous but modified. With the center at velocity *v*, the circle must be modified into the combination of two ellipses, one for the forward direction and one for the rearward as in Figures 5-9 and 5-10

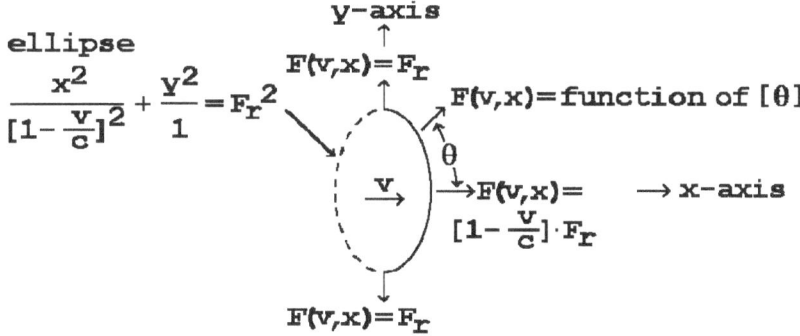

Figure 5-9
The Wave Resultant in the Forward Direction When the Center is at Velocity v

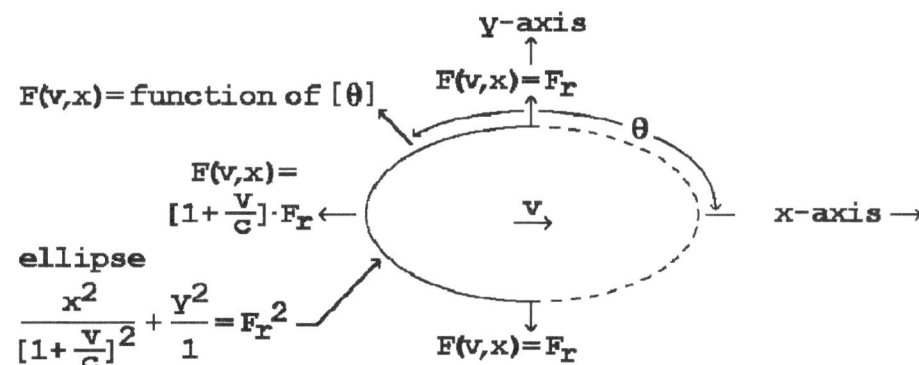

Figure 5-10
The Wave Resultant in the Rearward Direction When the Center is at Velocity v
The equations of these two ellipses as given in the above figures can be generalized as

(5-21) $\dfrac{x^2}{W^2} + y^2 = Fr^2$

where: W = 1-^{V}/$_C$ for +90° ≧ θ ≧ -90°
and
W = 1+^{V}/$_C$ for +90° ≦ θ ≦ +270°

Changing equation *5-21* from its rectangular coordinates (x, y) to polar coordinates in the variables (R, θ)

(5-22) $\dfrac{R^2 \cdot \text{Cos}^2(\theta)}{W^2} + R^2 \cdot \text{Sin}^2(\theta) = Fr^2$

is obtained, and solving for the radius, R, the result is

(5-23) $R = F_r \cdot \left[\dfrac{\text{Cos}^2\theta}{W^2} + \text{Sin}^2\theta \right]^{-\frac{1}{2}}$

Using equation *5-23* to express $F(v, x)$, as defined in Figures 5-9 and 5-10, in terms of the direction angle, θ,

(5-24)

$F(v,x) = F_r \cdot \left[\dfrac{\text{Cos}^2\theta}{W^2} + \text{Sin}^2\theta \right]^{-\frac{1}{2}}$

where: W = 1-^{V}/$_C$ for +90° ≧ θ ≧ -90°
and
W = 1+^{V}/$_C$ for +90° ≦ θ ≦ +270°

The analysis returns now to the overall situation per Figure 5-11, below, which is the same as the *velocity=0* case of Figure 5-3 except that: the charges are now in motion, *F(x)* is *F(v,x)*, *dF(x)* is *dF(v,x)* and angle θ is defined in the figure (and is functionally the same as in the above Figures 5-9 and 5-10).

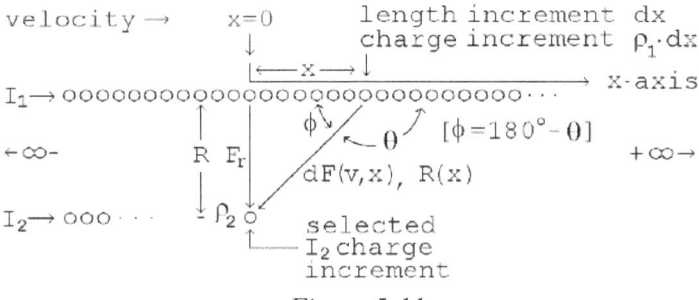

Figure 5-11

The following trigonometric relationships should be noted.

(5-25) Cos(180°-θ) = -Cosθ   Cos^2(180°-θ) = Cos^{2}θ
Cos(-θ) = Cosθ   Sin^2(180°-θ) = Sin^{2}θ

The integration performed before, for the case in which the velocity was zero, was of equation *5-14*, first line of which is repeated below.

(5-14) $\uparrow F_E = \displaystyle\int_{-\infty}^{0} \uparrow dF(x) + \int_{0}^{+\infty} \uparrow dF(x)$

CASES 1,2 AND 5

The Magnetic Force (The Charges In Motion)

In that expression FE is identical to FT because the velocity is zero so that $FM = 0$. Now, to deal with the charges in motion and non-zero velocity a new function, $f(v,x)$, is now defined:

$$(5\text{-}26) \qquad f(v,x) = \frac{dF(v,x)}{dF(x)}$$

so that

$$(5\text{-}27) \quad dF(v,x) = f(v,x) \cdot dF(x)$$

and this function will be integrated in the expression of equation *5-14* by substituting $dF(v,x)$ per equation *5-27* for $dF(x)$. The result will be FT, the total force at velocity v, rather than FE, the static case force.

To proceed, from the definition of $f(v,x)$ per equation *5-26*:

$(5\text{-}28) \quad f(v,x) = $ [Cases 1, 2, 5]

$$= \frac{\left[\begin{array}{c}\text{Wave Resultant} \\ \text{in direction } \theta \\ \text{at velocity } v\end{array}\right]}{\left[\begin{array}{c}\text{Wave Resultant} \\ \text{in direction } \theta \\ \text{at } v=0\end{array}\right]} \cdot \frac{\left[\begin{array}{c}\text{Center Resultant} \\ \text{in direction} \\ [\_90°\text{-}\theta] \text{ at } v\end{array}\right]}{\left[\begin{array}{c}\text{Center Resultant} \\ \text{in direction} \\ [\_90°\text{-}\theta] \text{ at } v=0\end{array}\right]} + \left[\begin{array}{c}F_{fc} \text{ Wve} \\ \text{toward} \\ \theta \\ \text{at } v\end{array}\right] - \left[\begin{array}{c}F_{fc} \text{ Ctr} \\ \text{toward} \\ 180°\text{-}\theta \\ \text{at } v\end{array}\right]$$

＊ At zero velocity there is no F_{fc} at all.

$$= \frac{\left[\dfrac{\cos^2\theta}{A^2} + \sin^2\theta\right]^{-\frac{1}{2}}}{1} \cdot \frac{\left[\dfrac{\cos^2\theta}{B^2} + \sin^2\theta\right]^{-\frac{1}{2}}}{1} - \left[\begin{array}{c}C \cdot \cos\theta - \\ D \cdot \cos\theta\end{array}\right]^{\text{＊＊}}$$

＊＊ Per equation 5-26, $f(v,x)$ is a measure of relative effects and does not include F_r.

Equations 5-25 have already been applied, and:

$\quad$ A = wave amplitude function of velocity
$\qquad$ = $[1 - v/c]$ forward and $[1 + v/c]$ rearward

$\quad$ B = center amplitude function of velocity
$\qquad$ = $[1 - v/c]$ forward and $[1 + v/c]$ rearward

$\quad$ C = wave F_{fc} function of velocity
$\qquad$ = $[+v/c]$ forward and $[-v/c]$ rearward

$\quad$ D = center F_{fc} function of velocity
$\qquad$ = $[+v/c]$ forward and $[-v/c]$ rearward.

Using equation *5-9* for $R(x)$ and the right triangle geometry of Figure 5-11, equation *5-28* becomes

(5-29) $f(v,x) =$ [Cases 1, 2, 5]

$$= \left[\frac{x^2/R(x)^2}{A^2} + \frac{R^2}{R(x)^2}\right]^{-\frac{1}{2}} \cdot \left[\frac{x^2/R(x)^2}{B^2} + \frac{R^2}{R(x)^2}\right]^{-\frac{1}{2}} + (C + D) \cdot \frac{x}{R(x)}$$

$$= \frac{A \cdot B \cdot (x^2 + R^2)}{[x^4 + (A^2 + B^2) \cdot R^2 \cdot x^2 + A^2 \cdot B^2 \cdot R^4]^{\frac{1}{2}}} + \frac{(C + D) \cdot x}{(x^2 + R^2)^{\frac{1}{2}}}$$

values of *A, B, C,* and *D* for which are given in the following Figure 5-12 for Cases *1, 2* and *5*.

Referring back to equation *5-13*, which separates the entire range to be calculated into two ranges, *-∞ to 0, [<0]*, and *0 to +∞, [>0]*, the Range column in the Figure 5-12 refers to those two ranges, the two ranges integrated separately because of the different values of *A*, *B*, *C* and *D* in the two ranges per the figure.

Case	Range	A	B	C	D
1 →	<0	$[1-v/c]$	$[1+v/c]$	$+v/c$	$-v/c$
→	>0	$[1+v/c]$	$[1-v/c]$	$-v/c$	$+v/c$
2 →	<0	$[1-v/c]$	$[1-v/c]$	$+v/c$	$+v/c$
←	>0	$[1+v/c]$	$[1+v/c]$	$-v/c$	$-v/c$
5 →	<0	$[1-v/c]$	1	$+v/c$	0
.	>0	$[1+v/c]$	1	$-v/c$	0

Figure 5-12

See Appendix D, *Integration Details for Magnetic Effect Calculations, Part 2*.

The expression to be integrated now is

(5-30)

$$\uparrow F_T = \int_{-\infty}^{0} \uparrow dF(v,x) + \int_{0}^{+\infty} \uparrow dF(v,x)$$

$$= \int_{-\infty}^{0} \uparrow f(v,x) \cdot dF(x) + \int_{0}^{+\infty} \uparrow f(v,x) \cdot dF(x)$$

$$= \int_{-\infty}^{0} \left[\left[\frac{A \cdot B \cdot (x^2 + R^2)}{[x^4 + (A^2 + B^2) \cdot R^2 \cdot x^2 + A^2 \cdot B^2 \cdot R^4]^{\frac{1}{2}}} + \cdots\right.\right.$$

$$\left.\left.\cdots + \frac{(C + D) \cdot x}{(x^2 + R^2)^{\frac{1}{2}}}\right] \cdot \left[F_r \cdot \frac{R^3}{[x^2 + R^2]^{1\frac{1}{2}}}\right]\right] \cdot dx + \cdots$$

$$\cdots + \int_{0}^{+\infty} \left[\begin{array}{c}\text{The same above entire}\\\text{expression a second}\\\text{time}\end{array}\right] \cdot dx$$

The integration and evaluation are at Appendix D, *Integration Details for Magnetic Effect Calculations*.

The result of that integration is for each of the 2 ranges

$$(5\text{-}31) \quad \uparrow F_T = R \cdot F_r \cdot [A \cdot B + C + D]_{>0} + R \cdot F_r \cdot [A \cdot B + C + D]_{<0}$$
$$= 2 \cdot R \cdot F_r \cdot [A \cdot B + C + D]$$

which is the result from the static case multiplied in each range by an expression that is the effect of velocity on the centers and their propagated waves. Per equation *5-20* the static force per equation *5-14* must now be subtracted from the new total force, F_T at velocity v per equation *5-31*, to obtain the net magnetic force, F_M.

As equation *5-12* the horizontal forces net to zero.

$$(5\text{-}32) \quad \overrightarrow{F_T} = 0$$

The final step in the calculation is to evaluate equation *5-31* for each of cases 1, 2 and 5 by inserting the values of A, B, C and D from the above Figure 5-12, and then determining F_M from the above two equations. The results are tabulated in Figure 5-13.

Case	Range	$A \cdot B + C + D$
1 $\rightarrow$ $\rightarrow$	<0	$[1 + v/c] \cdot [1 - v/c] - v/c + v/c = [1 - v^2/c^2]$
	>0	$[1 + v/c] \cdot [1 - v/c] - v/c + v/c = [1 - v^2/c^2]$
		Sum $= 2 \cdot [1 - v^2/c^2]$
		$\uparrow F_T = 2 \cdot R \cdot F_r \cdot [1 - v^2/c^2]$
		$\uparrow F_M = F_T - F_E$
		$= F_T - 2 \cdot R \cdot F_r$
		$= -2 \cdot R \cdot F_r \cdot [v^2/c^2]$
		(attraction)
2 $\rightarrow$ $\leftarrow$	<0	$[1 - v/c] \cdot [1 - v/c] + v/c + v/c = [1 + v^2/c^2]$
	>0	$[1 + v/c] \cdot [1 + v/c] - v/c - v/c = [1 + v^2/c^2]$
		Sum $= 2 \cdot [1 + v^2/c^2]$
		$\uparrow F_T = 2 \cdot R \cdot F_r \cdot [1 + v^2/c^2]$
		$\uparrow F_M = F_T - F_E$
		$= F_T - 2 \cdot R \cdot F_r$
		$= +2 \cdot R \cdot F_r \cdot [v^2/c^2]$
		(repulsion)
5 $\rightarrow$ 0	<0	$[1 - v/c] \cdot [1] + v/c + 0 \quad = [1]$
	>0	$[1 + v/c] \cdot [1] - v/c - 0 \quad = [1]$
		Sum $= [2]$
		$\uparrow F_T = 2 \cdot R \cdot F_r$
		$\uparrow F_M = F_T - F_E$
		$= F_T - 2 \cdot R \cdot F_r$
		$= 0$
		(no effect)

Figure 5-13

The above results agree exactly in direction with the force, F_M, for the three cases: 1, 2, and 5 of Figure 5-1. They agree exactly in magnitude with the force, F_M, per the earlier derivation from traditional 20th Century physics that the ratio of the magnetic force to the electrostatic force is $[v^2/c^2]$, equation *5-19*. Here, however, the magnetic effect is <u>derived</u> from the characteristics and behavior of *Spherical-Centers-of-Oscillation*. As stated earlier, magnetic field is merely the effect of changes in the electrostatic field effect due to changes in the oscillations of the centers and in their propagated waves caused by the centers being in motion rather than at rest.

CASES 3 AND 4

Cases 3 & 4 treat the two currents perpendicular to each other. The magnetic effect between perpendicular currents comes about because each of the two perpendicular currents has a component that is parallel to a component of the other. To derive the force of interaction between two perpendicular currents from the already obtained forces between parallel currents the procedure is as follows (See Figure 5-14 below).

Step (1) - Each of the two orthogonal currents, I_1 and I_2, is resolved into two components, one to the left and one to the right. Thus, I_1 is the resultant of I_{1L} and I_{1R} and similarly for I_2.

(5-32)
$$I_{1R} = I_{1L} = \tfrac{1}{2}\cdot\sqrt{2}\cdot I_1 \qquad\qquad I_{1L} \text{ attracts } I_{2R}$$
$$I_{2R} = I_{2L} = \tfrac{1}{2}\cdot\sqrt{2}\cdot I_2 \qquad\qquad I_{1R} \text{ repels } I_{2L}$$

Step (2) - I_{2R} is attracted by I_{1L} (Case 1) in the amount F_A using the form of equation *5-3* [F_M per unit length].

(5-33)
$$F_A = \frac{\mu}{2\cdot\pi}\cdot\frac{I_{1L}\cdot I_{2R}}{R_c} = \frac{\mu}{2\cdot\pi}\cdot\frac{\left[\tfrac{1}{2}\sqrt{2}\cdot I_1\right]\cdot\left[\tfrac{1}{2}\sqrt{2}\cdot I_2\right]}{R_c} = \frac{\mu}{2\cdot\pi}\cdot\frac{I_1\cdot I_2}{2\cdot R_c}$$

Similarly, I_{2L} is attracted by I_{1R} in the amount F_B.

(5-34)
$$F_B = \frac{\mu}{2\cdot\pi}\cdot\frac{I_{1R}\cdot I_{2L}}{R_c} = \frac{\mu}{2\cdot\pi}\cdot\frac{\left[\tfrac{1}{2}\sqrt{2}\cdot I_1\right]\cdot\left[\tfrac{1}{2}\sqrt{2}\cdot I_2\right]}{R_c} = \frac{\mu}{2\cdot\pi}\cdot\frac{I_1\cdot I_2}{2\cdot R_c}$$

These two forces are depicted in Figure 5-15, below

Step (3) - The resultant of the two forces, F_A and F_B, is F_m, the net magnetic force, and its magnitude per equation 5-35, next page, and its direction per Figure 5-15 are correct for Case 3.

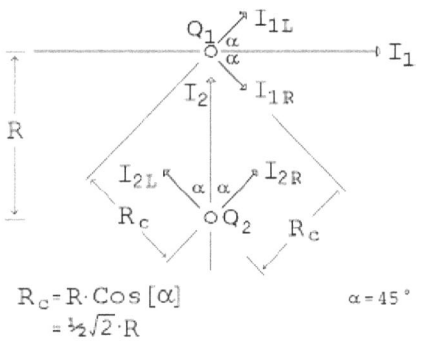

$$R_c = R\cdot Cos[\alpha]$$
$$= \tfrac{1}{2}\sqrt{2}\cdot R \qquad\qquad \alpha = 45°$$

Figure 5-14

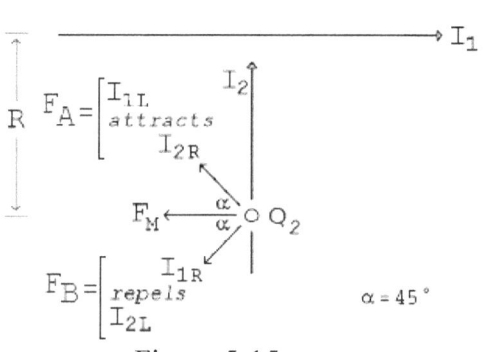

$\alpha = 45°$

Figure 5-15

(5-35)

$$F_M = \sqrt{2} \cdot F_A = \sqrt{2} \cdot F_B = \sqrt{2} \cdot \frac{\mu}{2 \cdot \pi} \cdot \frac{I_1 \cdot I_2}{2 \cdot R_c} = \frac{\mu}{2 \cdot \pi} \cdot \frac{I_1 \cdot I_2}{R}$$

[From Figure 5-14 $R_c = \frac{1}{2} \cdot \sqrt{2} \cdot R$].

If the direction of I_2 is reversed Case 4 is obtained and a review of the above analysis will show that the resulting magnitude of F_M is the same as for Case 3 while the direction of F_M is opposite to that of Case 3.

Thus the magnetic force in Cases 3 and 4 is demonstrated, which completes the derivation of magnetic field from *Spherical-Centers-of-Oscillation* considerations. As with Coulomb's Law and electrostatics, Ampere's Law and magnetostatics are now moved from the realm of empirical results to derived results, results derived from the origin of the universe and its implications.

The Electro-Magnetic Action (Varying Charge Velocity)

When the charges, the *Spherical-Centers-of-Oscillation*, are in motion at various speeds in various directions, they always exhibit the electrostatic Coulomb behavior at each "still" instant of their motion. And, when they are so in motion they always exhibit the magnetic Ampere behavior at each "momentary constant velocity" instant of their motion.

When the motion involves changing speed and / or changing direction the effect is a "stream" of the individual states exhibited at each instant of the motion, each successive state being in effect an imprint on the *Propagated Outward Flow*, of Figure 4-3 *"The Center's Propagation as Observed from At-Rest"*.

The "stream" is analogous to a motion picture projection of successive frames of a "moving picture" each frame slightly different from its predecessor. But, whereas the motion picture presents discrete "frames" the "stream" of states from electric charges is smooth and continuous and radially outward in all directions.

In electronic communications information is carried as variations, modulation, in the frequency or amplitude of a flat, smooth unchanging oscillation, the carrier wave. In the case of the stream of the individual states exhibited at each instant of the motion of an electric charge the "carrier wave" is the *Propagated Outward Flow* and the "information carried" modulation of the carrier is the succession of the various forms of the center's propagations. The variations in the electric charge's *Propagated Outward Flow* are a modulation of what its *Propagated Outward Flow* would be if the electric charge were not in motion.

That stream of flow modulated by a succession of the various states of motion of the source electric charges, or rather the actual modulation itself, appears to us as what we term Electro-Magnetic Waves: light, radio, television, and various communications.

61

$\longrightarrow$

The Action of Matter: Matter Waves

The Matter Wave Problem

In the early 20th Century it was proposed, in the absence of any knowledge of *Spherical-Centers-of-Oscillation*, that perhaps matter, which was accepted as being particle in nature might sometimes exhibit wave behavior. It was hypothesized that the wave aspect of a particle of matter should have a wavelength, λ_{mw}, of

(6-1)
$$\lambda_{mw} = \frac{h}{particle\ momentum} = \frac{h}{m \cdot v}$$

This was soon verified by the obtaining of electron diffraction patterns whose observed wavelengths corresponded well enough with the prediction. At that point one would think that the duality of matter was enough established that extensive further investigation of matter waves would have resulted. But that was not the case and the reason was a fundamental problem that could not be overcome – the matter wave frequency.

If one reasons that the kinetic energy of the particle of matter should correspond to its matter wave frequency, f_{mw}, as

(6-2)
$$f_{mw} = \frac{W_k}{h} = \frac{\frac{1}{2} \cdot m \cdot v^2}{h}$$

then the velocity of the matter wave is

(6-3)
$$v_{mw} = \lambda_{mw} \cdot f_{mw} = \left[\frac{h}{m \cdot v} \right] \cdot \left[\frac{\frac{1}{2} \cdot m \cdot v^2}{h} \right] = \frac{1}{2} \cdot v$$

which states that the matter wave moves at one half the speed of the particle. That is obviously absurd as they must move together each being merely an alternative aspect of the same real entity.

It is no help in resolving this difficulty if relativistic mass is used (as it should be in any case) since the same mass appears in both numerator and denominator of equation *6-3* where they simply cancel out.

It is also no help to hypothesize that it is the total energy, not just the kinetic energy, that yields the matter wave. Such an attempt attributes a matter wave to a particle at rest. It also gives the resulting matter wave velocity as c^2/v which has the matter wave racing ahead of its particle.

It was the inability to resolve this problem that led to the loss of interest in matter waves and essentially the end of further inquiry with regard to the wave aspect of matter.

Resolution of the Matter Wave Problem

If instead of kinetic energy one uses energy in kinetic form, $m_v \cdot v^2$, as developed in Section 4, *The Action of Matter: Motion and Relativity*, equation *4-13* the problem of the matter wave frequency is resolved. The traditional view of kinetic energy as the energy increase due to motion is not valid as a description of the processes taking place.

Using mass- and energy-in-kinetic-form to obtain the frequency of the matter wave proceeds as follows.

(6-4)
$$f_{mw} = \frac{m_v \cdot v^2}{h}$$
[equation *6-2*, but using W_v, equation *4-13*, energy-in-kinetic-form, for W_k, kinetic energy]

Using this result for matter wave frequency and using the same relativistic mass, m_v, in equation *6-5* for the matter wavelength the velocity of the matter wave then is

(6-5)
$$v_{mw} = f_{mw} \cdot \lambda_{mw} = \left[\frac{m_v \cdot v^2}{h} \right] \cdot \left[\frac{h}{m_v \cdot v} \right] = v$$

and the wave is traveling with and as the particle. On that basis the wave aspect of matter is established both experimentally and theoretically.

Matter Waves and Spherical Centers of Oscillation

The matter wave traveling right along with the particle is like a kind of standing wave relative to the particle. A standing wave can be thought of as the sum result of two waves traveling in opposite directions through each other. If the frequencies and wavelengths are different then their interaction produces a new frequency called a "beat". The development of the beat is as follows.

The two waves are

(6-6) Wave #1 $= A \cdot \mathrm{Sin}(2\pi f_1 t)$

Wave #2 $= A \cdot \mathrm{Sin}(2\pi f_2 t)$

and the sum is

(6-7) Wave Sum $= A \cdot \mathrm{Sin}\left[2\pi f_1 t\right] + A \cdot \mathrm{Sin}\left[2\pi f_2 t\right]$

which by using a trigonometric equivalence can be arranged as

$$\mathrm{Wave\,Sum} = 2A \cdot \mathrm{Sin}\left[2\pi \frac{f_1 + f_2}{2} t\right] \cdot \mathrm{Cos}\left[2\pi \frac{f_1 - f_2}{2}\right]$$

The cosine term frequency ½· *[f1-f2]* difference, is smaller than the sine term sum ½· *[f1+f2]*. If the expression is viewed as the higher frequency sine portion with the rest of the expression being the amplitude, as in equation *6-8*, then

(6-8)
$$\mathrm{Wave\,Sum} = \left[2A \cdot \mathrm{Cos}\left[2\pi \frac{f_1 - f_2}{2} t\right]\right] \cdot \mathrm{Sin}\left[2\pi \frac{f_1 + f_2}{2} t\right]$$

$$= \left[\mathrm{Varying\ Amplitude}\right] \cdot \mathrm{Sin}\left[2\pi \frac{f_1 + f_2}{2} t\right]$$

The wave form appears as in Figure 6-1, below.

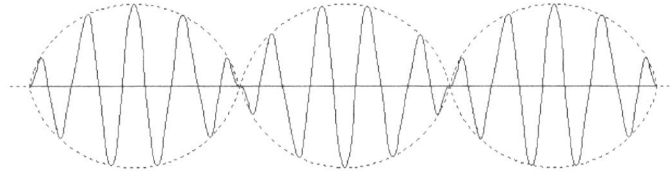

Figure 6-1

The solid-line curve in Figure 6-1 is the overall wave form. The dotted line, the *envelope*, is the varying amplitude. The overall wave form exhibits in the varying amplitude a periodic variation called the *beat*. The beat is real, not merely an appearance. For example two sound tones heard simultaneously produce an audible beat that one can hear. It is by listening to the beat that one tunes a piano or other musical instrument.

Matter waves are the beat that results from the *Spherical-Center-of-Oscillation*'s forward and rearward oscillations interacting with each other. This develops as follows. For a center in motion at velocity v, per Figure 4-3

$$(6-9) \quad \begin{aligned} \lambda_{fwd} &= \lambda_v \cdot (1 - {}^v\!/c) & f_{fwd} &= {}^c\!/\lambda_{fwd} \\ \lambda_{rwd} &= \lambda_v \cdot (1 + {}^v\!/c) & f_{rwd} &= {}^c\!/\lambda_{rwd} \end{aligned}$$

The beat frequency, using the "Varying Amplitude" portion of equation *6-8*, substituting f_{fwd} for f_1 and f_{rwd} for f_2, and then using equation *6-9*, is

$$(6-10) \quad f_{beat} = \frac{1}{2}\left[f_{fwd} - f_{rwd}\right] = \frac{1}{2}\left[\frac{c}{\lambda_v\left[1-\frac{v}{c}\right]} - \frac{c}{\lambda_v\left[1+\frac{v}{c}\right]}\right]$$

$$= \frac{c}{2\cdot\lambda_v}\cdot\left[\frac{\left[1+\frac{v}{c}\right]-\left[1-\frac{v}{c}\right]}{\left[1-\frac{v}{c}\right]^2}\right] = \frac{v}{\lambda_v}\cdot\left[\frac{1}{\left[1-\frac{v}{c}\right]^2}\right]$$

$$\lambda_{beat} = \frac{c}{f_{beat}} = \lambda_v\cdot\frac{c}{v}\left[1-\frac{v^2}{c^2}\right]$$

Substitute Eqn $4-2$

$$\lambda_v = \lambda_r\cdot\frac{1}{\left[1-\frac{v^2}{c^2}\right]^{\frac{1}{2}}}$$

$$= \left[\lambda_r\frac{1}{\left[1-\frac{v^2}{c^2}\right]^{\frac{1}{2}}}\right]\cdot\frac{c}{v}\cdot\left[1-\frac{v^2}{c^2}\right]$$

Substitute per :

$$m\cdot c^2 = h\cdot f = h\cdot\frac{c}{\lambda} \rightarrow \lambda_r = \frac{h}{m_r\cdot c}$$

$$= \left[\frac{h}{m_r\cdot c}\right]\cdot\frac{c}{v}\cdot\left[\left[1-\frac{v^2}{c^2}\right]^{\frac{1}{2}}\right]$$

Substitute per Eqn $4-6$

$$m_v = m_r\cdot\frac{1}{\left[1-\frac{v^2}{c^2}\right]^{\frac{1}{2}}}$$

$$= \frac{h}{m_v\cdot v}$$

which is the matter wavelength as previously obtained per equation *6-1* (in which the mass must be relativistic mass, m_v, of course). Thus matter waves are the beat that results from the *Spherical-Center-of-Oscillation*'s forward and rearward oscillations interacting with each other.

A moving center-of-oscillation as "seen" by an external observer appears as the waves propagated by the center in his direction appear. But, if one could, somehow, actually "see" the center itself pulsating as it does, the situation would be different. The interaction of the forward and rearward oscillations, which produce a beat at the matter wave frequency, are real. The effect is as follows (repeating the form of equations *6-6* through *6-8*, which were for any general oscillation, but now using the oscillations of a center-of-oscillation in motion).

(6-11) $\text{Forward Wave} = A \cdot \left[1 + \text{Sin}(2\pi f_1 t)\right]$

$\text{Rearward Wave} = A \cdot \left[1 + \text{Sin}(2\pi f_2 t)\right]$

```
[Note: 1 - cos(x)  ≡ 1 + cos(180° - x)
                   ≡ 1 + sin[90°-(180° - x)]
                   ≡ 1 + sin(x - 90°)
     and the 90° phase is irrelevant, of course.]
```

and the sum is

(6-12) $\text{Wave Sum} = A \cdot \left[2 + \text{Sin}\left[2\pi f_1 t\right] + \text{Sin}\left[2\pi f_2 t\right]\right]$

Which again by using a trigonometric equivalence can be arranged as

$$\text{Wave Sum} = 2A + 2A \cdot \text{Sin}\left[2\pi \frac{f_1 + f_2}{2} t\right] \cdot \text{Cos}\left[2\pi \frac{f_1 - f_2}{2}\right]$$

The cosine term is at a lesser frequency than the sine term. If the expression for the wave sum is viewed as the (higher frequency) sine portion with the rest of the expression being the amplitude, as in equation *6-13*, then

(6-13)
$$\text{Wave Sum} = 2A \cdot \left[1 + \text{Cos}\left[2\pi \frac{f_1 - f_2}{2} t\right]\right] \cdot \text{Sin}\left[2\pi \frac{f_1 + f_2}{2} t\right]$$

$$= 2A \cdot \left[\begin{array}{c}1 + \text{cosine form of} \\ \text{Varying Amplitude}\end{array}\right] \cdot \text{Sin}\left[2\pi \frac{f_1 + f_2}{2} t\right]$$

In the case of a *Spherical-Center-of-Oscillation* $f_1 = f_{fwd}$ and $f_2 = f_{rwd}$. Likewise, A is U_C, the center average amplitude, the oscillation being of the form $U_C \cdot [1 - Cos(2\pi \cdot f \cdot t)]$ as before, equation *1-16*.

The wave form appears as in Figure 6-2, below, for the forward-rearward interaction and the matter wave beat of the center's pulsation as it would be "seen" from the side relative to its direction of motion.

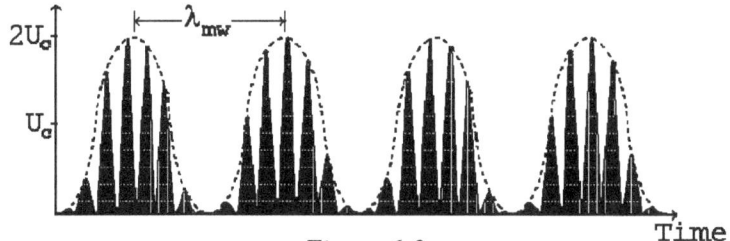

Figure 6-2
The Forward-Rearward Pulsation of a Center in Motion
Which is the Matter Wave

Matter Waves and Electron Orbits In Atoms

By the early 20th Century it had become clear that atoms consist of a positively charged minute nucleus surrounded mostly by empty space except for a moderate number of negative, equally charged, minute electrons in orbits around the nucleus the orbital configuration maintained by a balance of Coulomb attraction and centrifugal force. It also had become clear that there were only a small number of discrete orbits that were stable, that is orbits that supported a continuous cyclical electron path around the nucleus.

Intriguingly it had also been found that the stable orbits are only those whose orbital path length is exactly an integer multiple of the orbiting electron's matter wave length. One would have expected that such a significant correlation would have led to extensive further investigation of matter waves and of the correlation. However, the then unsolved problem of the matter wave frequency [first page of this section] resulted in general neglect of matter waves.

It also resulted in the invention of an alternative so far as electron orbits are concerned. The statement that the orbital electron stable path length is an integer multiple of the electron's matter wave length, equation *6-14*,

(6-14) $\begin{array}{ll} \text{Orbital} \\ \text{Path} \end{array}$ = $\begin{array}{ll} \text{Matter Wave Length} \\ \text{Integer Multiple} \end{array}$

$$2\pi \cdot R \quad = \quad n \cdot \frac{h}{m \cdot v} \quad = \quad n \cdot \lambda_{mw} \qquad n = 1, 2, \ldots$$

was algebraically modified [by switching the location of the *2π* and the *m·v*] to state that the orbiting electron's angular momentum occurred in only integer multiples of a fundamental quantity [Planck Constant over 2π], i.e. are "quantized", equation 6-15.

(6-15) $\begin{array}{ll} \text{Orbital} \\ \text{Angular} \\ \text{Momentum} \end{array}$ = "Quantized"

$$m \cdot v \cdot R \quad = \quad n \cdot \frac{h}{2\pi} \qquad\qquad n = 1, \ 2, \ldots$$

There is no cause, no mechanism that requires the orbital angular momentum to be "quantized". But that the stable orbits are only those whose orbital path length is exactly an integer multiple of the orbiting electron's matter wave length is due to a specific behavior of the Spherical-Centers-of-Oscillation *as follows.*

Taking the simple case of the Hydrogen atom with its single proton nucleus and single orbital electron the Coulomb attraction by the positive atomic nucleus on the negative orbiting electron is not a smooth continuous action. Rather it is the result of an on-going stream of pulses at the rate of the frequency of oscillation of the proton *Spherical-Center-of-Oscillation* that is per equation *2-6b* a frequency of *2.268,731,818·10²³* hz when the proton is at rest.

For the orbit to be stable it must be the same for each pass, pass after pass. If each pass includes exactly an integer number of the orbital electron's matter wave lengths then each pass has exactly the same set of Coulomb force pulses acting in each orbital pass. But if, for example, the orbital path length contains only $^{9}/_{10}$ of a matter wave length, $^{9}/_{10}$ of the matter wave period, then the next pass will contain the missing $^{1}/_{10}$ of the matter wave length or wave period plus $^{8}/_{10}$ of the next, and so on. The matter wave being sinusoidal in form, the successive orbital passes will be all different, in Figure 6-2.

It is this behavior which operatively causes the "stable orbits", and only those orbits, to be stable. It has nothing to do with angular momentum nor quantization of angular momentum.

How Electrons Are Forced Into Stable Orbits

With the vast amount of *Propagated Outward Flow* from myriad *Spherical-Centers-of-Oscillation* orbital electrons are continuously buffeted. How are specific stable orbit paths enforced? To analyze and quantify the deviations in the variable quantities involved, the radius, R, and the electron orbital velocity, v, will be expressed in terms of the orbit number, n, the number of matter wavelengths in the orbital path. That requires obtaining expressions for them that do not include any other variables.

That quantity, n, will here be deemed to be a continuous variable so that the R and v expressed in terms of n can be continuously variable and able to address locations between stable orbits, not merely the discrete amounts at the stable orbits.

The balance of forces for stability in a circular orbit requires

(6-16)   Centrifugal Force = Centripetal Force

$$\frac{m \cdot v^2}{R} = \frac{q^2}{4\pi \cdot \varepsilon \cdot R^2}$$

$$R = \frac{q^2}{4\pi \cdot \varepsilon \cdot m \cdot v^2}$$

(6-17)   Orbit Path Length = n · Matter Wavelength

$$2\pi \cdot R = n \cdot \frac{h}{m \cdot v}$$

$$2\pi \left[\frac{q^2}{4\pi \cdot \varepsilon \cdot m \cdot v^2} \right] = n \cdot \frac{h}{m \cdot v} \qquad \text{[Substitute } R]$$

$$v = \frac{q^2}{2\pi \cdot \varepsilon \cdot n \cdot h} \qquad \text{[Solve for } v]$$

$$v \propto \frac{1}{n}$$

(6-18) $$R = \frac{q^2}{4\pi \cdot \varepsilon \cdot m \cdot v^2} \propto \frac{q^2}{4\pi \cdot \varepsilon \cdot m \cdot \left[\dfrac{1}{n} \right]^2} \qquad \text{[Substitute } 6\text{-}17]$$

$$R \propto n^2$$

In those terms the variation of the required centripetal force for a circular orbit as n varies is

(6-19)
$$F_{Centripetal} = \frac{m \cdot v^2}{R} \propto \frac{\left[\frac{1}{n} \right]^2}{n^2} = \frac{1}{n^4}$$

With constant charge the only variable in the expression for the Coulomb force is R in the denominator and is proportional to n^4. Therefore

(6-20) $F_{Coulomb} \propto 1/n^4$

Thus the normal Coulomb force always provides the exact value of $F_{centripetal}$ required for a stable circular orbit.

The numerator of the Coulomb force expression is q^2. The variation from the force it exerts in the stable orbits depends on the ratio of the orbital path length, $2\pi \cdot R$, to the matter wavelength, $h/m \cdot v$. If that ratio is an integer then the behavior is the normal stable orbit Coulomb force.

If that ratio is not an integer then the force is *quasi-stable Coulomb*, as if the effective charge were modified as follows.

(6-20)

$$\text{Coulomb Force Numerator} \propto \frac{\text{Orbit Length}}{\lambda_{mw}}$$

$$\propto \frac{2\pi \cdot \mathbf{R}}{h/m\mathbf{v}} = \frac{2\pi \cdot \mathbf{R} \cdot m \cdot \mathbf{v}}{h}$$

$$\propto n^2 \cdot [1/n] = n$$

$$\text{Coulomb Force Denominator} \propto \mathbf{R}^2 \propto n^4$$

and the overall *quasi-stable Coulomb* force then varies as

(6-21)

$$F_{Quasi-Coulomb} = \frac{\text{Numerator}}{\text{Denominator}} \propto \frac{n}{n^4} = 1/n^3$$

The ratio of the quasi-Coulomb force to the normal Coulomb force then varies as

(6-22)

$$\frac{F_{Quasi-Coulomb}}{F_{Normal\ Coulomb}} = \frac{1/n^3}{1/n^4} = n$$

This means that for values of n somewhat larger than that of the next lower stable orbit integer value the actual Coulomb force acting, $F_{Quasi-Coulomb}$, is too large. For values of n somewhat below the stable orbit integer value the actual Coulomb force acting, $F_{Quasi-Coulomb}$, is too small.

Those results mean that:

- *Outside or above the stable orbit* integer value of orbit n the excessive values of $F_{Quasi-Coulomb}$ have the *net effect of moving the electron path inward*. The inward force produces an inward acceleration that is greater than the amount to produce a circular orbit. The excess acceleration produces inward electron velocity. (The inward $F_{Quasi-Coulomb}$ is greater than the outward "centrifugal force".)

- *Inside or below the stable orbit* integer value of orbit n the insufficient values of $F_{Quasi-Coulomb}$ have the *net effect of moving the electron path outward*. The inward force produces an inward acceleration that is less than the circular orbit amount. The deficiency produces less than circular motion, a net outward motion effect. (The inward $F_{Quasi-Coulomb}$ is less than the outward "centrifugal force".)

The overall effect is to force the electrons into stable orbits as Figure 6-3.

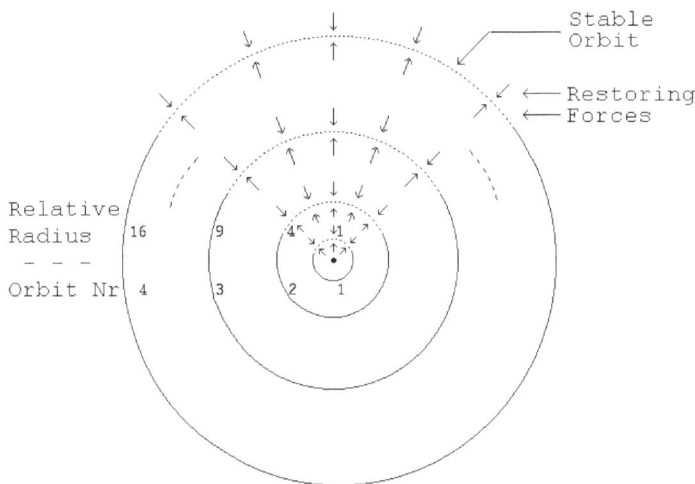

Figure 6-3
The Orbital Electrons Forced Into Integer Matter Wavelength Orbits

The Electrons' Transition Paths Between Stable Orbits

The above Figure 6-3 depicts the status when the orbital electrons are all in their lowest [least energy] orbits. When the outermost of those orbital locations is not occupied and the electron that should be in that position is in an excessively higher orbital location the action of the restoring forces is to direct that electron inward on an orbital transition path to fill the unoccupied position. That happens as follows.

The absence of an electron in the unoccupied position means that the positive electron-attracting field of the atom's positive nucleus is slightly un-offset by the orbital electrons' negative charges. With all of the lowest orbits filled the atom overall presents an electrically neutral status as viewed from outside, but with the outer electron missing that presentation is slightly of inner positive charge as viewed from the excessively higher orbital location electron.

That extra centrally directed attraction curves the pattern of restoring forces of Figure 6-3 to that of Figure 6-4, below.

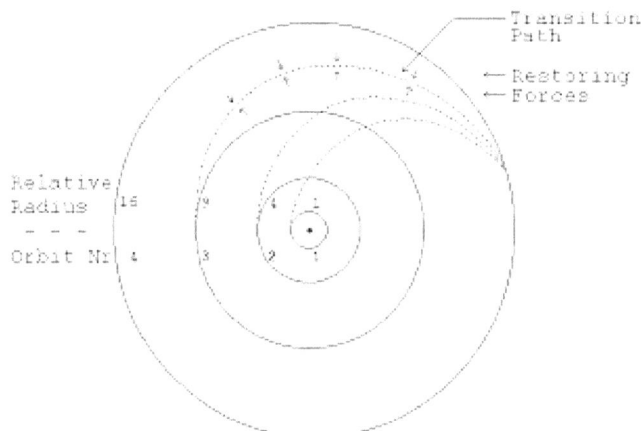

Figure 6-4
The Electrons Orbit Transition Paths

70

That drives the excessively higher orbital location electron inward to fill the empty location.

Any vacant location in the lowest energy positions of the orbital electron structure is automatically filled from above by this directing of the restoring forces. That is how an outer electron "knows" that there is a space that it can and should move into and that is how it follows the correct path to get there.

From any point in an outer orbit there is one specific path to each of the inner orbits of that outer orbit. Such paths, which involve inward motion of the electron in transition between stable orbits, have at each point in their path the correct inward motion to compensate for the deviation of the value there of $F_{Quasi-Coulomb}$ from what the normal Coulomb force should be at that point.

The electron velocity must vary smoothly from the stable velocity of the initial outer orbit through a period of increase and ending in the stable velocity of the final orbit. To do that without a discontinuity the variation must be in the form of a half cycle cosine. That is attested to by the sinusoidal nature of the *E-M* radiated photon. There can only be one such path that correctly compensates between any particular pair of initial and final orbits.

On either side of such a path the transition path restoring forces act just as for the stable orbits. The restoring forces arise because the stable orbit restoring forces will not allow locations between stable orbits.

Multi-Electron Atoms' Orbital Electrons Structure

Finally the question arises: what is the allocation of electrons to the stable orbits in multi-electron atoms and what impels the electrons into that structure?

In effect the orbital electron extends a distance of $\frac{1}{2} \cdot \lambda_{mw}$ forward and rearward of its instantaneous location. The space that the matter wave occupies is like a long straight narrow tube tangential to the electron's location on its orbital path.

There are three constraints that govern the behavior of the orbital electrons:

(1) The orbital path length must be an integral number of matter wavelengths, as already developed;

(2) The electrons being all of the same charge magnitude and polarity, tend to repel each other to a spacing equally apart subject to the common central attraction of the oppositely charged nucleus;

(3) The electron spacing along the orbital path must be such that the $\frac{1}{2} \cdot \lambda_{mw}$ extension of the electron in space forward and rearward of its current position does not interfere with the space correspondingly occupied by any of the other electrons.

Of course, in addition there are the obvious constraints that the number of electrons in orbit must be the same as the number of equivalent positive charges in the nucleus because the atom is overall electrically neutral and that the electron orbits and the electron positions in the orbits must be such that they do not collide nor otherwise interfere with each other.

The 20th Century physics model of the orbital electron arrangements is that the electrons are arranged in "shells" [as if in spherical surfaces] designated *n=1*, *n=2*, etc., and that there is space for a maximum of: *2* electrons in the *n=1* shell, *8* electrons in the *n=2* shell, *18* electrons in the *n=3* shell, and so on. Those dispositions are correct, but the rules used to determine them, a set of four "Orbital Quantum Numbers", provide no mechanism, no cause for the behavior.

The orbital electron allocation to orbits and arrangement is enforced by the requirement of accommodating the space that each orbiting electron's matter wave occupies, as follows.

Applying the constraints to the innermost *n=1* shell where the orbital path length is $n \cdot \lambda_{mw} = 1 \cdot \lambda_{mw}$ there is only space for *2* electrons in the orbital plane [see Figure 6-4 and equation *(6-23)*, below]. In the figure the second electron is depicted located as close to the first electron as possible without their matter wave extensions in space interfering with each other. Introduction of a third electron into that orbit in that plane would involve spacing that would disrupt the particles and the orbit.

(6-23)   For the n = 1 shell the orbital path length, is one
wavelength, $2\pi \cdot R = \lambda_{mw}$. Then from Figure 6-4, below:

$$Tan(\Phi) = \frac{\frac{1}{2} \cdot \lambda_{mw}}{R} = \frac{\frac{1}{2} \cdot 2\pi \cdot R}{R} = \pi$$

$\Phi = 72.34^{\circ}$

Electron Space = $360^{\circ}/_{2 \cdot \Phi} = 2.49 \Rightarrow 2$ electrons

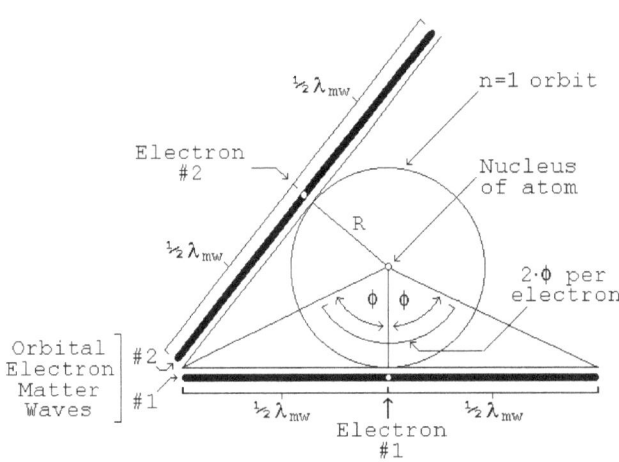

Figure 6-5
Electrons in n=1 Shell

Considering adding more orbital planes, the situation is like a sword dance where a number of dancers whirl and turn, each flashing a pair of swords [matter wave occupied space], while avoiding any casualties among the dancers. The dancers' spacing, paths and timing must be such that while their swords slash at each others' paths they do so only when the dancer in that path and his extended swords is out of the way.

If a plane tilted relative to the above first orbital plane is introduced in the *n=1* shell its first electron will interfere with the prior two regardless of the tilt. Imagining in Figure 6-4, above, that the paper is folded along the line from the nucleus to where the two matter waves just meet the fold tilts one electron's orbital plane relative to the other but does not change the interference of the two. Thus, in terms of the angles in Figure 6-4, a second orbital plane tilted at an angle of *φ = 72.34°* or more would seem to fit.

However, the electron in that second orbital plane, starting at $\Phi = 72.34º$ above one of the points of intersection with the first plane could travel only the distance $[180°-2\cdot\Phi] = 35.32º$ before being within $\Phi = 72.34º$ of the other side of the orbit, the other point of intersection of the planes. During that $35.32°$ the pair of electrons in the original plane have not had the necessary travel, $\Phi = 72.34º$, to clear their matter wave extensions in space from the common points of intersection of the two orbital planes.

Therefore, the *n=1* shell can only contain *one orbital plane* with only *one orbit* having *two equally spaced electrons*. Any additional content would involve the matter waves of the electrons interfering with each other.

For the *n=2* shell the "sword dance" becomes more complex. Clearly, from the above, the first two *n=2* electrons can readily share an orbit as in the *n=1* case. In fact, calculation analogous to equation *(6-23)* but for the *n=2* case shows that three electrons could fit in one *n=2* orbital plane. That calculation is as follows.

(6-24)   For the n = 2 "shell" the orbital path length
        the circular path circumference, is two matter
        wavelengths, $2\pi\cdot R = 2\cdot\lambda_{mw}$.

$$Tan(\Phi) = \frac{\frac{1}{2}\cdot\lambda_{mw}}{R} = \frac{\frac{1}{2}\cdot\pi\cdot R}{R} = \pi/2$$

$\Phi = 57.52º$

Electron Space = $360º/2\cdot\Phi = 3.13 \Rightarrow$ 3 electrons

However, the fit is close and more overall equidistant spacing of the electrons is achieved with the third electron occupying a new orbital plane tilted to the first as develops below.

How many such tilted planes can be accommodated at the *n=2* level in total ? The shell can accommodate three such planes at $\theta = 60º$ relative tilts. This limit is set by $\Phi_{n=2} = 57.52º$ per equation *6-24*. Four planes tilted at $\theta = 45º$ would be too close. The three planes have a common axis of intersection on which are the two points that all three of the orbits have in common (Figure 6-6).

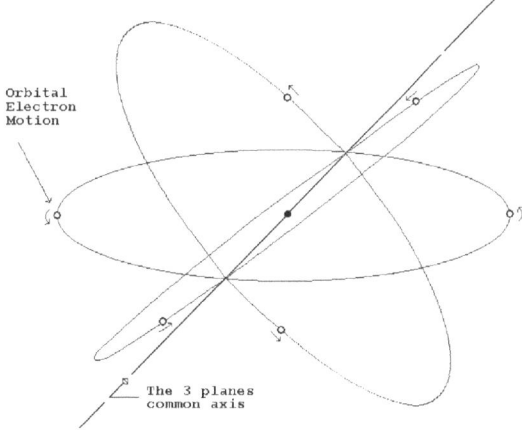

Figure 6-6
Three Orbital Planes and Relative Tilts, n=2 Shell

The six electrons (two per each of three orbital planes tilted at $60°$ relative to each other) pass through those two common points at $\varphi = 360°/6 = 60°$ intervals (equidistant spacing). With $\Phi = 57.52°$ there is just enough travel between successive electrons for each electron to clear the area before the next one starts arriving.

Can any more electrons fit in this shell ? Yes, two more in another orbital plane perpendicular to the common axis of the other three orbital planes. This new orbit intersects each of the other three successively at $\Theta = 60°$ intervals. The two electrons in each such intersected plane are spaced $180°$ apart. An electron passing such an intersection with one of the first three planes $60°$ after one of that plane's two electron's has passed and taking $60°$ to clear the intersection would have cleared the requisite $60°$ ahead of the other electron of that plane.

Two such electrons $180°$ apart can be accommodated.

Overall, therefore the $n=2$ shell can fit *eight electrons – two in each of the three common axis planes* plus *two more in the perpendicular plane*.

For $n=3$ the situation becomes considerably more complex. Now the separation angle is $\Phi_{n=3} = 46.32°$. The reasoning as for $n=2$, above, indicates that the shell can still accommodate only three orbital planes intersecting on a common axis, each plane having two electrons in orbit $180°$ apart with the one more plane perpendicular to the common axis of the other three planes. In other words, for $n=3$ the shell appears able to only accommodate the same orbital structure as does the $n=2$ shell. This is in fact the case.

More precisely, the $n=3$ shell so functions until full in that form. Additional electrons for higher z atoms then start filling the $n=4$ shell. Then, the electric field of those outer $n=4$ electrons becomes sufficient to modify the orbital structure situation and possibilities of the inner $n=3$ shell. The $n=3$ shell then can accommodate the expected five orbital planes on a common axis, each with two electrons, in addition to the already filled $n=2$ type structure.

For higher n the same kind of effect of outer on inner shell modifies the structure, the $n=5$ shell filling partly before the $n=4$ shell is completely filled and that partial outer shell's field then modifying the inner shell's structure.

It is the complex fitting of the space occupied by the orbital electron matter waves into the available integer-matter-wavelength orbital shells that determines the orbital electrons' arrangement structure. That structure is summarized in the table of Figure 6-7, below. The table, arranged so as to directly correspond to the "quantum numbers" system of 20th Century physics shows what those quantum numbers actually represent.

The entire structural effect is the result of the matter waves of the orbital electrons and the restrictions that their space requirements impose on the system.

```
"Quantum
Number"              Orbital Structure

   n          The shell's orbital path length is "n"
              matter wavelengths long.

                   n = 1, 2, 3, …

   l           The number of "sets" in a particular
              "shell" equals [l + 1].

                   l = 0, 1, … n-1

              A "set" consists of orbital planes of
              orbits of the same length, tilted at
              equal angles relative to each other,
              and sharing the same common axis about
              which tilted.

   m_l        The "index number" of any particular
              orbital plane in any particular "set"
              of orbital planes.

                   m_l = +[ l ], +[ l − 1], … 0, -1, … -[ l ]

              The total number of such orbital planes
              in the "set" is

              [ 2·l + 1], always odd.

   m_S        Each individual orbital plane can
              accommodate 2 electrons equally spaced.

                   m_S = -½ and +½ [for the 1^st and 2^nd
                        electrons of the plane].
```

Figure 6-7

$\longrightarrow$

The Action of Matter: Gravitation

INTRODUCTION

This presentation of the gravitational behavior of *Spherical-Centers-of-Oscillation* is in two parts.

-The first, *The Conceptual Aspect*, deals with

[a] the defects in Einstein's General Theory of Relativity namely: that its proposed cause or mechanism of gravitation is incomplete and that its tests can validate only the tested effects not the proposed cause or mechanism, and

[b] conceptually how the *Spherical-Centers-of-Oscillation* produce a complete gravitational cause and mechanism in a manner consistent with the rest of physics.

- The second is the analytical and mathematical *Derivation of Newton's Law of Gravitation* from first principles of *Spherical-Centers-of-Oscillation* and further *Proof that Inertial Mass and Gravitational Mass Are Identical i.e. the Same.*

THE CONCEPTUAL ASPECT

The problem with generally accepted theories that are nevertheless in error lies not primarily with their error but rather that they are an impediment to the consideration of alternative more valid theories. The classic example of this is the case of Ptolemy's geocentric theory of the motion of the solar system's planets.

The Ptolemaic [Earth centered] system accounted for the observed motion of the planets by hypothesizing a motion variant in the form of cycles and epicycles imposed on the apparent general orbital path around the Earth. That system successfully accounted for the planets' motions and successfully predicted eclipses, retrograde motion, and alignments although it needed small adjustments from time to time over the approximately 1,500 years of its dominant rule.   But, it lacked a valid cause or mechanism.

A physics theory consists of known experimentally experienced behaviors or effects and a physical explanation of the cause or mechanism producing those observed effects.

- That a theory passes test predictions of its *effects* does not validate its proposed *mechanism*.

[*e.g.* Ptolemy: geocentric planetary system].

- An *incomplete* proposed cause or *mechanism* leaves its theory unvalidated.

[*e.g.* Einstein: gravitational mass curves space].

[How does it curve space, what mechanism ?]
[What is "space" that it can be "curved" ?]

77

The behavior of gravitation is well known, described by Newton's Law of Gravitation. But what gravitational mass is, how gravitational behavior comes about, what in material reality produces the effects of gravitation, is little understood.

Experience shows that everything has a cause and that those causes are themselves the results of precedent causes, *ad infinitum*. Defining and comprehending the causality or mechanism operating to produce an observed behavior is essential to understanding or explaining it.

The comprehensive explanation of the cause and mechanism of gravitation as derived from the origin of the universe is the Modern Newtonian Model of Gravitation. Its development consists of the following seven steps. Each step results in new "hard" facts generated directly from prior "hard" facts. The development does not contain nor rely on opinion. Consequently, while it is deemed a "model" it is an exact factual description of what it treats.

1 – How the universe's particles of matter came into existence.

2 – How they came to be propagating an outward flow.

3 – The reservoir supply for the substance of the outward flow.

4 – The speed of the outward flow.

5 – A particle's flow encountering another particle slows its outward flow.

6 – The outward flow has momentum.

7 – Gravitation is the momentum reaction to outward flow slowing.

STEP 1 – HOW THE UNIVERSE'S PARTICLES OF MATTER CAME INTO EXISTENCE

This has been fully developed in this book's Section 1, *The Origin of Matter: Its Cause*.

STEP 2 – HOW THE MATTER PARTICLES CAME TO BE PROPAGATING AN OUTWARD FLOW

This has been fully developed in this book's Section 2, *The Behavior of Matter: Its Form*, under the heading *The Flow from the Spherical-Centers-of-Oscillation*.

STEP 3 – THE RESERVOIR SUPPLY FOR THE SUBSTANCE OF THE OUTWARD FLOW

This has been fully developed in this book's Section 2, *The Behavior of Matter: Its Form*, under the sub-heading *The Particle Core's Propagated Outward Flow*.

STEP 4 – THE SPEED OF THE OUTWARD FLOW

This has been fully developed in this book's Section 2, *The Behavior of Matter: Its Form*, under the sub-heading *The Speed of the Flow – The Speed of Light*.

STEP 5 – A PARTICLE'S FLOW ENCOUNTERING ANOTHER PARTICLE SLOWS ITS OUTWARD FLOW

In a universe of the myriad particles, *Spherical-Centers-of-Oscillation*, resulting from the Big Bang, each of those particles propagating its own *Propagated Outward*

Flow radially in all directions, there are many instances of the flow from one particle [the "source" particle] encountering, running into, the outward-flow-propagating-center core of another particle [the "encountered" particle]. Such "source" particle flows are inverse square reduced in magnitude the farther that their wave front has traveled from its "source".

The flow behavior is analogous to that of an electric transmission line where the rate of travel of an oscillation down the line is determined by the time it takes to build up the electric current for each oscillation cycle through each infinitesimal increment of the line's distributed series inductance [L_p] and to build up the electric potential for each oscillation cycle on each infinitesimal increment [C_p] of the line's distributed shunt capacitance. The transmission line speed of flow is determined by the well-established relationship

$$(7\text{-}1) \qquad \text{Speed} = \frac{1}{\sqrt{L_p \cdot C_p}}$$

For *Spherical-Centers-of-Oscillation* propagating oscillating flow the factor determining the speed of propagation is the time required to build up the flow amount for each oscillation cycle through each infinitesimal increment of the flow's μ_0 and the flow's potential for each oscillation cycle on each infinitesimal increment of the flow's ε_0. But, in radially outward propagating particle's flow, the flow amount is inverse square spread out and the potential likewise, both in exactly the same proportion as its μ_0 and ε_0. The ratio of the flow amount to its μ_0 and of its flow potential to its ε_0 remains constant, and so likewise the speed, radially outward, of its propagation, c.

Upon encountering another particle that arriving flow's μ and ε (scalar not vector) (much inverse square reduced) combine with the (full magnitude) μ_0 and ε_0 in the new outgoing propagation of the encountered center, the $\mu_0 + \mu$ sum and the $\varepsilon_0 + \varepsilon$ sum each being larger values. The result is that that "encountered" particle's new outward flow is slowed relative to its natural speed. That is, its speed of flow is determined by a combination of the parameters μ and ε larger than its flow's otherwise natural values. The speed of flow is determined by the well-established relationship

$$(7\text{-}2) \qquad \text{Speed} = \frac{1}{\sqrt{\mu \cdot \varepsilon}}$$

STEP 6 – *THE OUTWARD FLOW HAS MOMENTUM*

The oscillating substance, *Medium*, of each of the myriad particles is its mass. There is no other place or thing to be the mass of those particles. Therefore the propagating outward flow has momentum, the inherent effect of the product of mass, inherent in the substance of the flow, and the flow's velocity.

In the absence of other effects the outward flow is naturally radially outward. While the outward flow effectively transmits pulses of momentum outward in its [1 - Cosine] oscillation, the core source of that flow is experiencing radially <u>inward</u> equal but opposite pulses of momentum in accordance with Newton's third law of motion. In effect the core source is under reaction compression. Because that effect is radially uniform it produces no net affect on the particle.

79

STEP 7 – GRAVITATION IS THE MOMENTUM REACTION TO OUTWARD FLOW SLOWING.

The incoming flow from a distant "source" particle having the effect of slowing the speed of the "encountered" particle's outward propagated flow causes that "encountered" particle's outward flow to have less momentum than if it were not slowed, again momentum being the product of mass and velocity.

Therefore the Newton's Third Law reaction to that reduced outward flow momentum, reaction back on the "encountered" particle, is smaller than otherwise. That effect takes place on the side of the "encountered particle" facing toward the "source" particle from which the slowing - causing flow came.

[Newton's Third Law, every action has an equal opposite reaction, is valid because without it masses could self-accelerate in violation of conserving energy.]

But, on the opposite side of the "encountered" particle no such slowing of its outward propagated flow is present so that the outward flow there has the full natural momentum and the Newton's Third Law reaction on the particle on that side is the full natural amount. Consequently, the "encountered" particle experiencing its usual full momentum reaction back on itself on its side opposite that facing the incoming flow from the "source" but experiencing reduced reaction back on itself on its side facing the incoming flow from the "source", experiences a net momentum reaction toward the "source" particle from which the slowing-causing flow came.

Thus the particle experiences *[1 - Cosine]* pulses of momentum increase toward the "source" gravitationally attracting particle which constitute the gravitational acceleration.

DERIVATION OF NEWTON'S LAW OF GRAVITATION

In Section 2 a statement of the derived gravitational acceleration was obtained as equation *2-17*, repeated below.

(2-17)
$$\Delta v = c \cdot \frac{\delta^2}{d^2} \quad \text{per cycle of } f_{source}$$

a quite pure, precise and direct statement of the operation of gravitation. It states that gravitation is a function of the speed of light, *c*, and the inverse square law, in the context of the oscillation frequency, f_S, corresponding to the attracting, source body's mass. It should be noted that equation *2-17* is exact without involving a constant of proportionality such as *G*.

The equation *2-17* result can also be obtained directly from consideration of solely how slowing is caused by μ and ε, which demonstrates that the cause of gravitation is the slowing of wave propagation presented just above. That is as follows.

For the *Medium* of the *Propagated Outward Flow* at the instant of its propagation from its source center responding to its own μ_0 and ε_0, the value of those two are constant at what we term their free space values. Those values are inverse square reduced as the medium carrying them propagates outward from their source center-of-oscillation. (As discussed in the prior section, the speed of wave propagation remains the same because the waves are also inverse square reduced in amplitude.)

(7-3)   (1) At distance δ from the center of the source center,
the first place where the propagated medium
appears and where its concentration is greatest,
the values of μ and ε are the free space values:

$$\mu = \mu_0 \qquad \text{and} \qquad \varepsilon = \varepsilon_0$$

(2) Per the inverse square law, the values at distance "d"
from the center of the source center are:

$$\mu(d) = \mu_0 \cdot \frac{\delta^2}{d^2} \qquad \text{and} \qquad \varepsilon(d) = \varepsilon_0 \cdot \frac{\delta^2}{d^2}$$

Then, the overall net effective values when flowing medium from a distant center passes through the outward propagation of an encountered center are

(7-4)

$$\mu_{net} = \left[\mu_0 + \mu_0 \cdot \frac{\delta^2}{d^2}\right] = \mu_0 \cdot \left[1 + \frac{\delta^2}{d^2}\right]$$

$$\varepsilon_{net} = \left[\varepsilon_0 + \varepsilon_0 \cdot \frac{\delta^2}{d^2}\right] = \varepsilon_0 \cdot \left[1 + \frac{\delta^2}{d^2}\right]$$

The resulting net speed of propagation is, then

(7-5)

$$c_{net} = \frac{1}{\left[\mu_{net} \cdot \varepsilon_{net}\right]^{\frac{1}{2}}} = \frac{1}{\left[1 + \frac{\delta^2}{d^2}\right] \cdot \left[\mu_0 \cdot \varepsilon_0\right]^{\frac{1}{2}}}$$

$$= \frac{c}{\left[1 + \frac{\delta^2}{d^2}\right]} = \frac{d^2}{d^2 + \delta^2} \cdot c$$

and the amount of the slowing is

(7-6) $\Delta c = c - c_{net}$

$$= c \cdot \left[1 - \frac{d^2}{d^2 + \delta^2}\right]$$

$$= c \cdot \frac{\delta^2}{d^2 + \delta^2}$$

$$= c \cdot \frac{\delta^2}{d^2} \qquad \text{[}d^2 \text{ is much greater than } \delta^2\text{]}$$

so that

(7-7)

$$\Delta v = c \cdot \frac{\delta^2}{d^2} \qquad \begin{array}{l}\text{[the slowing, } \Delta c, \text{ equals} \\ \text{the velocity change, } \Delta v\text{]}\end{array}$$

which is identical to equation *2-17*, above.

Equation *(2-17)*, above, gives the gravitationally caused velocity change per cycle of the incoming gravitational wave field. The time rate of those velocity change increments, i.e. the gravitational acceleration, a_g, is Δv times the incoming wave's frequency, which is the source center's frequency, f_s.

$(7\text{-}8)$ $\quad a_g = \Delta v \cdot f_s$

$$= c \cdot \frac{\delta^2}{d^2} \cdot f_s$$

$$= c \cdot \frac{\delta^2}{d^2} \cdot \frac{m_s \cdot c^2}{h} \qquad [m_s = \text{the source center's mass;}$$

$$= G \cdot \frac{m_s}{d^2} \qquad \begin{array}{l} [\text{substituting } G \text{ per equation } 2\text{-}14 \text{ re} \\ \text{the definition of the Planck Length} \\ \text{and equation } 2\text{-}16 \text{ re definition of } \delta] \end{array}$$

$(7\text{-}9)$ $\quad F_g = a_g \cdot m_e$

$$= G \cdot \frac{m_s \cdot m_s}{d^2} \qquad [m_e \text{ is the encountered center's mass.}]$$

which is Newton's Law of Gravitation.

As with Coulomb's Law and inertial mass, as developed in Section 3, here gravitation, gravitational mass, and Newton's Law of gravitation cease to be mere empirically valid observations becoming instead requisite behavior aspects of natural reality derived from *Spherical-Centers-of-Oscillation* fundamentals.

INERTIAL MASS AND GRAVITATIONAL MASS ARE IDENTICAL *i.e.* THE SAME

From equation *(7-8)* it is clear that only m_s operates in the process of gravitation. That is, m_s in equation *(7-8)* is a gravitational mass and the gravitational acceleration is independent of m_e. Equation *(7-9)* on the other hand is merely a statement of Newton's 2nd Law. The new [relative to equation *(7-8)*] mass in equation *(7-9)*, m_e, is the inertial mass of Newton's law.

But, the overall action is mutual. The "encountered" center gravitationally attracts the "source" center at the same time as the "source" center gravitationally attracts the "encountered" center. Therefore, the quantities, m_s and m_e, are, and operate simultaneously, in both roles – "source" and "encountered", as both types of mass – inertial and gravitational. That means that inertial mass and gravitational mass are identical as follows.

Given two gravitationally attracting bodies, *#1* and *#2*, the force with which *#1* attracts *#2* must equal that with which *#2* attracts *#1* [Newton's 3rd Law of Motion]. That is

$(7\text{-}10)$ Using: "Grav" = the gravitation constant [normally "G"]
Body #1 has inertial mass = i
and gravitational mass = g
Body #2 has inertial mass = I
and gravitational mass = G
Then:
Force$_{grav}$ #1⇨#2 = Force$_{grav}$ #2⇨#1

$$\text{"Grav"} \cdot \frac{g \cdot I}{d^2} = \text{"Grav"} \cdot \frac{G \cdot i}{d^2}$$

$$g \cdot I = G \cdot i$$

(7-11)   If: Inertial Mass ≡ Gravitational Mass

              I = G   and   i = g

    Then: g·I = G·i   [of equation 7-10 above]

    Can Be: g·G = G·g   and   i·I = I·i

    Which is obviously true.

whereby proving that:

The inertial mass and the gravitational mass are identical.

[That conclusion, has long been thought by modern physicists to be the case, and has been indicated by the most sophisticated measurements, but has been beyond proof because of the lack of understanding of gravitation.]

CONCLUSION

The physical effects that we refer to as gravitational mass and gravitation and their behavior in Newton's Law of Gravitation are all properly understood as actions of *Spherical-Centers-of-Oscillation* via their *Propagated Outward Flow.*

The comprehensive development and set of derivations and proofs of:

- The Origin of Matter.
- Coulomb's Law
- The Lorentz Transforms
- Ampere's Law
- Matter Waves
- Orbital Electrons Behavior
- Gravitation

demonstrate the overall validity of the *Spherical-Centers-of-Oscillation* and their *Propagated Outward Flow* description of material reality which must replace the various "field" theories of modern physics as well as Einstein's General Theory of Relativity treatment of gravitation all of which fail because of their lack of developed causes and mechanisms.

$\longrightarrow$

Appendices

Appendix A-1

The Neutron

The fundamental, basic and most simple particles, the proton and the electron and their anti-particles, are developed in the preceding Section 2, *The Behavior of Matter: Its Form*. The other usually stable particles, the atomic nuclei and the neutron, are combinations of those basic particles as developed in this Appendix A – Particles.

The evidence that the neutron is a combination of an electron and a proton is overwhelming.

- Unlike the case with atomic nuclei, where the presence of multiple protons and their mutual electrostatic repulsion makes the nucleus tend to fly apart except for the nuclear binding energy [treated in Appendix A-2], an electron and a proton would tend to bind together in mutual electrostatic attraction. No binding energy or mass deficiency would be needed for an electron - proton combination.

- This correlates with the neutron mass, which exceeds the sum of the masses of the hypothesized components, a proton and an electron, by $0.000,839,854$ amu (more than the mass of an electron). The neutron has in this sense a negative mass deficiency or binding energy, a mass excess. One would expect this since the act of combining a proton and an electron should also include at least some of the energy of their mutual attraction.

- Because of the negative binding energy one would expect the neutron to be unstable. While it is stable in a stable atomic nucleus, where it is affected by its overall nuclear environment, free on its own it readily decays into a proton and an electron with a mean lifetime before decay of about 881.5 seconds.

- Of course the combination naturally yields the neutron's electrostatic neutrality.

The primary traditional objection to the concept stems from the matter wave wavelength of the electron. In that view the wavelength associated with the electron component of the proton-electron combination would be far too large and in direct contradiction to observed cross-sections and wavelengths.

However, that objection applies to a "bunch of grapes" concept of the two particles' combination – their, so to speak, sitting side by side like two peas in a pod. But if the two particles combine more intimately into a new neutron form their waves combine more intimately. Figure A-1-1, below, shows the combination of two oscillations at very different frequencies, the higher representing the proton and the lower, the electron of the proton - electron pair of which a neutron would be composed.

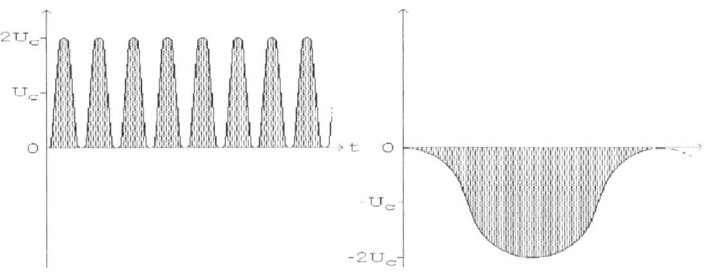

Figure A-1-1(a)
Proton & Electron Oscillations, Two Different Frequencies

As in Figure A-1-1(b), below while our eyes can perceive the longer wavelength in the combined wave form (the envelope), the actual oscillation is only at a wavelength essentially that of the shorter input wavelength. The electron's matter wave need not be a problem.

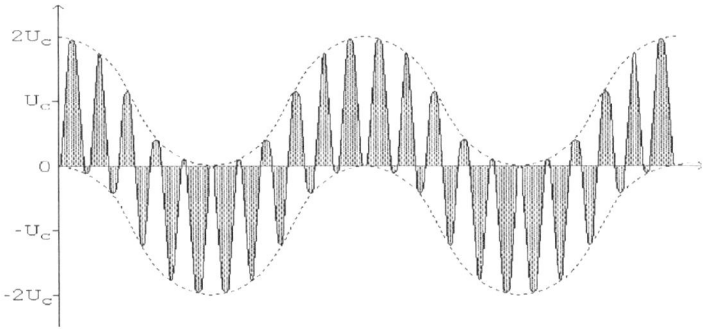

Figure A-1-1(b)
The Sum Oscillation, The Neutron

The *Spherical-Center-of-Oscillation* equation of the neutron, depicted in Figure A-1-1 above, is the sum of those equations of the electron and the proton.

$$(A-1-1) \quad U(_0n^1) = U_c \cdot \left[1 - Cos(2\pi f_p t)\right] - U_c \cdot \left[1 - Cos(2\pi f_e t)\right]$$
$$= U_c \cdot \left[Cos(2\pi f_e t - Cos(2\pi f_p t)\right]$$

The masses of the proton and electron the combination of which is the neutron are not their rest masses even though their combination in the neutron yields the neutron's rest mass. The component masses are the particles' relativistic masses at high velocity. This comes about as follows.

Since a neutron naturally decays into a proton and an electron those decay particles must be emitted at a velocity equal to or greater than their escape velocity. That is, because the proton and electron strongly mutually attract each other electrically, unless they separate at their mutual escape velocities they will immediately re-combine into a neutron.

Put another way, for a neutron to be formed from a proton and an electron the two must come together from the state of being mutually independent of each other. That means that they must mutually accelerate toward each other. In so doing they will each be at escape velocity and have the kinetic energy of that escape velocity at the moment of their combining into the new particle, the neutron.

The portion of the neutron's overall rest mass that corresponds to the component proton and electron's escape velocity kinetic energy is the neutron rest mass less the sum of the proton and the electron rest masses.

(A-1-2) Δm_n = $m_{neutron,\ rest}$ $-$ $[m_{proton,rest}$ $+$ $m_{electron,\ rest}]$

$\qquad$ = 1.008,664,904 - …

$\qquad\qquad$ … - [1.007,276,470 + 0.000,548,579,903]

$\qquad$ = 0.000,839,854 amu.

In the "classical" sense escape velocity refers to an object of some mass that is gravitationally bound to some other mass, for example a rocket to be launched from Earth. The force attracting the two objects, the rocket and the Earth, to each other acts on them equally in magnitude and opposite in direction. Consequently, momentums that are equal in magnitude and opposite in direction are imparted to them. Since momentum is the product of mass and velocity, when one object (Earth) is much more massive than the other (the rocket) it may be assumed with negligible error that it (the Earth) is not accelerated and its velocity is negligible. Then all of the kinetic energy is attributable solely to the rocket. That kinetic energy must be equal to the gravitational potential energy binding the rocket to the Earth (the two to each other) for the rocket to escape the Earth's gravitational pull.

However, in the case of a proton and an electron the assumption that only the particle of lesser mass is accelerated and that the other particle's kinetic energy is negligible is not valid. It is not that the electron escapes from the proton; they escape from each other. Or, it is not that the electron falls toward the proton; they fall toward each other. The kinetic energy of each is involved and the sum of the kinetic energies must equal or exceed the binding potential energy for the velocities to be at or in excess of escape velocity.

The analysis is as follows (where r is the closest separation between the escaping objects or particles).

(A-1-3) Gravitational $\qquad\qquad\qquad\qquad$ Electrostatic

$\quad$ Rocket [R] escapes from $\qquad\qquad$ Proton [p] and electron
$\quad$ from Earth [E] $\qquad\qquad\qquad\qquad\quad$ [e] escape each other

$\qquad\qquad\qquad\qquad$ (a) PE = Force·r

$$PE = \left[G \cdot \frac{m_R \cdot m_E}{r^2} \right] \cdot r \qquad\qquad PE = \left[\frac{q_p \cdot q_e}{4 \cdot \pi \cdot \varepsilon_0 \cdot r^2} \right] \cdot r$$

$\qquad\qquad$ (b) Final (escape) Kinetic Energy (KE)
$\qquad\qquad\qquad$ = Initial Potential energy (PE)

$\qquad KE_R = PE_{total} \qquad\qquad\qquad\qquad KE_p + KE_e = PE_{total}$

$$\tfrac{1}{2} \cdot m_R \cdot v_R{}^2 = G \cdot \frac{m_R \cdot m_E}{r} \qquad\qquad \text{No direct solution}$$

$\qquad\qquad\qquad\qquad\qquad\qquad\qquad$ A 2nd relationship is
$\qquad\qquad\qquad\qquad\qquad\qquad |P_p| = |-P_e|$ P is momentum

$$v_{R,esc} = \left[\frac{2 \cdot G \cdot m_E}{r} \right]^{1/2} \qquad\qquad \begin{array}{l} \text{The two relationships} \\ \text{must be simultaneously} \\ \text{solved for the velocities} \end{array}$$

For the gravitational case the escape velocity formulation does not involve the mass of the escaping object. In that sense it is independent of the relativistic mass increase with velocity. Furthermore, in the usual cases treating escape velocity of objects (rocketry and astronautics) the velocity never approaches magnitudes at which significant relativistic effects occur.

However, for the electrostatic case the escape velocity formulation must include the masses of the particles, which masses themselves can vary with their velocity. And, in the case of particles, velocities large enough to involve relativistic effects are likely to occur. Therefore, the electrostatic case must be treated relativistically. The simultaneous solution of the electrostatic case's two equations, kinetic energy and momentum, is as follows.

(A-1-4) Momentum

Magnitude of Proton   Magnitude of Electron
Relativistic Momentum = Relativistic Momentum

$$\frac{m_p}{\left[1-\frac{v_p^2}{c^2}\right]^{\frac{1}{2}}}\cdot v_p = \frac{m_e}{\left[1-\frac{v_e^2}{c^2}\right]^{\frac{1}{2}}}\cdot v_e \qquad m_p \text{ \& } m_e \text{ are rest masses}$$

Solving the above for v_p the following is obtained.

(A-1-5) [Momentum continued]

$$v_p = \frac{m_e\cdot v_e}{m_p\cdot\left[1-\frac{v_e^2}{c^2}\right]^{\frac{1}{2}}}\cdot\frac{1}{\left[1+\frac{m_e^2\cdot v_e^2}{c^2\cdot m_p^2\cdot\left[1-\frac{v_e^2}{c^2}\right]}\right]^{\frac{1}{2}}}$$

(A-1-6) Energy

Relativistic Energy [As Mass] Is Conserved

$$\left[\frac{KE_p+KE_e}{c^2}\right]_{gained} = \left[\frac{PE_{total}}{c^2}\right]_{lost}$$

$$\left[m_{p,v}-m_{p,rest}\right]+\left[m_{e,v}-m_{e,rest}\right]=m_n-\left[m_{p,rest}+m_{e,rest}\right]\equiv m_{n,\Delta}$$

$$[\frac{m_p}{\left[1-\frac{v_p^2}{c^2}\right]^{\frac{1}{2}}}-m_p]+[\frac{m_e}{\left[1-\frac{v_e^2}{c^2}\right]^{\frac{1}{2}}}-m_e]=m_{n,\Delta}$$

The above equations treat the excess of the neutron's rest mass above the sum of the rest mass of a proton plus that of an electron to be the relativistic KE gained by the two particles in approaching each other from infinite separation distance [per the concept of "escape velocity"].

The issue here is: how far apart are the proton and electron in their collision paths toward each other when they have the above kinetic masses, $m_{p,v}$ and $m_{e,v}$? For the calculations to be correct, that is for the hypothesis to be correct, their separation distance at that moment must be such that the two colliding particles are exactly at the moment of combining into the neutron. First the velocities, v_p and v_e, will be calculated by the simultaneous solution of equations *(A-1-4)* and *(A-1-5)*. Then the separation distance of the two particles at the moment of collision will be determined.

(A-1-7) Simultaneous Solution of A-1-4 With A-1-5

The expression for v_p from equation (4) is substituted for v_p in the denominator of the first term of the expression obtained in equation (5). The resulting expression has only v_e unknown and is solved for that value.

Rather than manipulating that expression a computer aided design program is used to calculate selected trial values of v_e until the correct result for $m_{n,\Delta}$ [$m_{n,\Delta} = m_n - m_{p,rest} - m_{e,rest}$] is obtained.

The results of that process are as follows.

(A-1-8) v_e = 275,370,263. m/s

$\qquad\qquad = 0.918,536,33 \cdot c$

$\qquad v_p$ = 379,350.6975 m/s

$\qquad\qquad = 0.001,265,378 \cdot c$

At those velocities the proton and the electron have total (relativistic) masses of

(A-1-9)

$$m_{e,total} = \frac{m_{e,rest}}{\left[1 - \dfrac{v_e^2}{c^2}\right]^{\frac{1}{2}}} = 2.529,490,15 \cdot m_{e,rest}$$

$$= 0.001,388,308,25 \text{ amu}$$

(A-1-10)

$$m_{p,total} = \frac{m_{p,rest}}{\left[1 - \dfrac{v_p^2}{c^2}\right]^{\frac{1}{2}}} = 1.000,000,80 \cdot m_{p,rest}$$

$$= 1.007,276,596 \text{ amu}$$

and their sum is the mass of the neutron.

(A-1-11) $m_{neutron} = m_{p,total} + m_{e,total}$

$$= 1.007,276,596 + 0.001,388,308,25$$

$$= 1.008,664,904 \text{ amu}$$

(These calculations assume that the component proton and electron are in a state of zero momentum and zero kinetic energy before being mutually accelerated into each other. It likewise assumes that the resulting neutron has zero kinetic energy and zero momentum because all the components' kinetic energy goes entirely into the neutron's rest mass and the two component's momentums are equal and opposite in direction netting to zero in combination. To the extent that the components do have initial kinetic energy and momentum then either the resulting neutron will have kinetic energy equal to the sum of the components' initial kinetic energies and momentum equal to the net of the two components' initial momenta or some of those quantities may appear in the form of some type of neutrino given off at the time the particles combine.

(Likewise, in describing the decay of a neutron into a proton and an electron, it was assumed that the neutron initially had zero kinetic energy and zero momentum. To the extent that that is not the case then some form of neutrino will account for the kinetic energy and net momentum not accounted for by the decay product proton and electron.)

THE REMAINING ISSUE IS:
HOW FAR APART ARE THE PROTON AND ELECTRON IN THEIR COLLISION PATHS TOWARD EACH OTHER WHEN THEY HAVE THE ABOVE KINETIC MASSES ?

Their separation distance at that moment must be such that the two colliding particles are exactly at the moment of combining into the neutron.

An initial calculation of that separation distance, r, is as follows.

(A-1-12)

$$[\text{Potential Energy}_{\text{As Mass}}] \equiv \frac{PE}{c^2} \text{ and must } = m_{n,\Delta}$$

$$\frac{PE}{c^2} = \frac{q_{proton} \cdot q_{electron}}{4\pi \cdot \varepsilon_0 \cdot r} \cdot \frac{1}{c^2} = [0.000,839,854 \text{ amu}] \cdot [^{kg}/_{amu}]$$

$$r = \frac{q_{proton} \cdot q_{electron}}{4\pi \cdot \varepsilon_0 \cdot c^2} = \frac{1}{[0.000,839,854 \text{ amu}] \cdot [^{kg}/_{amu}]}$$

$$r = 1.840,636,27 \cdot 10^{-15} \text{ meters}.$$

Some years ago experiments involving measurement of the scattering of charged particles by atomic nuclei, yielded an empirical formula for the approximate value of the radius of an atomic nucleus to be

(A-1-13) Radius $= [1.2 \cdot 10^{-15}] \cdot [\text{Atomic Mass Number}]$ meters

which formula would indicate that the proton radius (atomic mass number $A = 1$) is about $1.2 \cdot 10^{-15}$ meters.

The mass of the proton can be expressed as an equivalent energy, $m \cdot c^2$, and that as an equivalent frequency, $m \cdot c^2/h$, or an equivalent wavelength, $^h/_{m \cdot c}$. That wavelength (not a "matter wavelength") for the proton is

(A-1-14) $\lambda_p = 1.321,408,96 \cdot 10^{-15}$ meters

quite near to the empirical value for the proton radius from equation *(A-1-11)*.

Thus the initial calculation of the separation distance of the proton and electron when their kinetic masses are just correct for them to form a neutron, equations *(A-1-9)*, *(A-1-10)* and *(A-1-11)*, results in a separation distance of about *1½* proton radii or equivalent wavelengths, equation *(A-1-12)*. That uncorrected result is so close as to essentially validate that the neutron is a combination of a proton and an electron.

However, there is more.

The result at equation *(A-1-12)* must be corrected for a variation in the magnitude of the classical Coulomb interaction as the charges approach near to each other. The direction of the electrostatic effect of a charge is radial to the charge location. At great distances from a charge all of those radii in a local sample are such a small part of the total spherical Coulomb action that they are effectively parallel. But, near to the charge they all effectively diverge (as, of course, they actually do in all cases). That reduces the electrostatic force and requires the charges to approach each other more closely than to the distance calculated at equation *(A-1-12)* – in fact to a separation distance of λ_p exactly, within the limitations of the precision of our data. This develops as follows.

When the two charges are relatively near to each other there is less Coulomb effect because of the radial direction of the Coulomb effect action relative to the charges. Coulomb's law, expressed as potential energy as in equation *(A-1-12)*, above, now becomes as follows.

(A-1-15)

$$[\text{Potential Energy}_{\text{As Mass}}] = \frac{[\text{Reduction Factor}] \cdot \text{PE}}{c^2}$$

$$= [\text{Reduction Factor}] \cdot \frac{q_{\text{proton}} \cdot q_{\text{electron}}}{4\pi \cdot \varepsilon_0 \cdot r} \cdot \frac{1}{c^2}$$

$$\text{and must} = m_{n, \triangle} = [0.000,839,854 \text{ amu}] \cdot [^{kg}/_{amu}]$$

But, what is the formulation for the *Reduction Factor*?

For the analysis of the effect of the two charges being so near to each other that the radial divergence of the rays is significant the illustration and dimensions of Figure A-1-2, below, are used. In order to be useful the figure is greatly exaggerated, that is α, β, d and so forth are actually too minute to be seen in an unexaggerated figure.

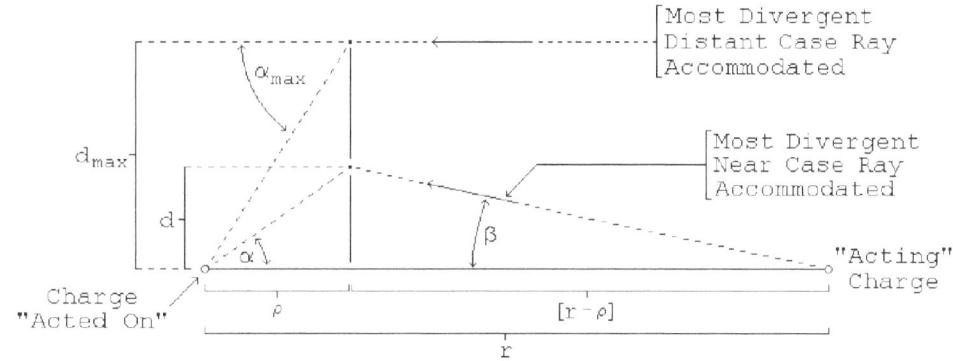

Figure A-1-2
Analysis of Case of Charges Close to Each Other

Even in the case of charges that are far apart, there is only one single ray that is a straight line from one charge to the other. All other rays of electric field must diverge at least minutely from that one straight ray. Therefore, because of the consistent behavior of the Coulomb Law for charges at a variety of separation distances, there is, in effect, a single constant angle of deviation that accommodates those of the divergent rays that enter into the effect. There must be some such angle which is essentially the same for all cases until the charges are close enough that the radial divergence affects the result. That angle is termed α_{max} in this development.

In terms of Figure A-1-2, for the case of charges near to each other, α_{max} must accommodate both β and α. When the charges are far apart β is essentially zero so that $\alpha_{max} = \alpha$. But, the maximum angle, α_{max}, all of which is available to α when the ray source is distant, must, when the ray source is near, account first for removing any ray divergence, β, with any remaining balance left for α. Therefore

(A-1-16) $\quad \alpha + \beta = \alpha_{max}$

(The quantity ρ is needed in order for the concept of α_{max} to have meaning; the angle is pointless without defining where it acts. For charges that are far apart α_{max} and ρ are of no significance. When near effects are operating ρ is at $r/2$, half-way between the charges. The concept of ρ is only included here for the initial purpose of presenting in the above Figure A-1-2 the comparison of the near and distant cases.)

The *Reduction Factor* depends upon the reduction of d (of Figure A-1-2) relative to d_{max}, that is the ratio d/d_{max} which quantity is developed as follows.

The angles α, α_{max}, and β are so small that their respective tangents equal their respective angles. Therefore, from the figure

(A-1-17)

$$\text{Tan}[\alpha_{max}] = \alpha_{max} = \frac{d_{max}}{\rho}$$

$$\text{Tan}[\beta_{max}] = \beta_{max} = \frac{d_{max}}{r - \rho}$$

$$\text{Tan}[\alpha] = \alpha = \frac{d}{\rho}$$

$$\text{Tan}[\beta] = \beta = \frac{d}{r - \rho}$$

From which

$$\alpha = \frac{d}{d_{max}} \cdot \alpha_{max}$$

$$\beta = \frac{d}{d_{max}} \cdot \beta_{max}$$

Then, substituting the above results into equation *(A-1-16)* the following is obtained.

(A-1-18) $\quad \alpha_{max} = \alpha + \beta$

$$= \frac{d}{d_{max}} \cdot \alpha_{max} + \frac{d}{d_{max}} \cdot \beta_{max}$$

From which

$$\frac{d}{d_{max}} = \frac{\alpha_{max}}{\alpha_{max} + \beta_{max}}$$

However, α_{max} is a constant quantity (from the consistent Coulomb behavior when the charges are far apart) as is d_{max}.

(A-1-19) $\quad \alpha_{max} = [\text{A Constant}] \cdot d_{max} \equiv \chi \cdot d_{max}$

Substituting for α_{max} of equation *(A-1-18)* with equation *(A-1-19)* and for β_{max} of equation *(A-1-18)* with β_{max} of equation *(A-1-17)* the *Reduction*

Factor sought for equation *(A-1-15)* is obtained. It is the d/d_{max} of equation *(A-1-20)*, below.

$$(A\text{-}1\text{-}20) \qquad \begin{bmatrix} \text{Reduction} \\ \text{Factor} \end{bmatrix} = \frac{d}{d_{max}} = \frac{\chi \cdot d_{max}}{\chi \cdot d_{max} + \dfrac{d_{max}}{r - \rho}}$$

$$= \frac{1}{1 + \dfrac{1}{\chi \cdot [r - \rho]}}$$

This *Reduction Factor* effect is also the cause of the *Lamb Shift*. The Lamb Shift is an extremely slight shifting to higher energy of the inner orbital energy levels of Hydrogen [Coulomb interaction at close separation as analyzed here]. That is, the Lamb Shift is greater as r is smaller. For that reason, it produces a detectable affect principally on the electrons of the inner orbital shells *[n = 1* or *n = 2]*.

The form of the effect is depicted graphically in Figure 3, below.

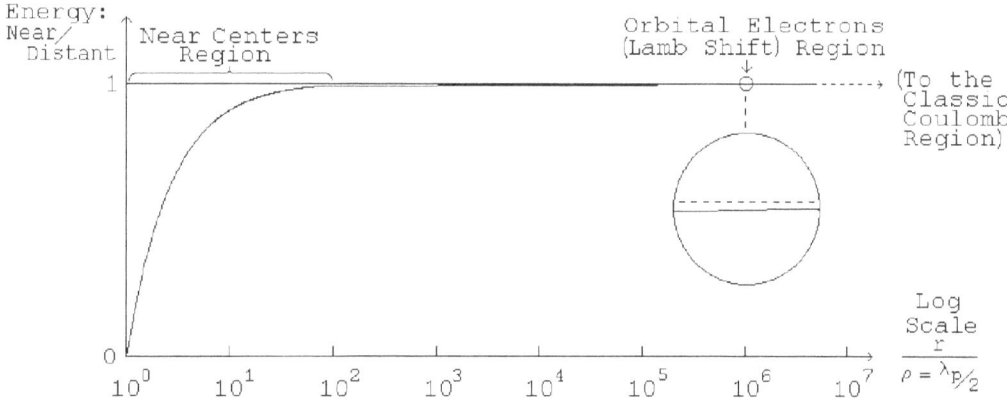

Figure A-1-3
Coulomb Effect Reduction Factor When Charges Are Near to Each Other

The Lamb Shift was attributed to "radiative coupling of the electron to the zero point fluctuation of the vacuum". What that means in plain language is as follows. Heisenberg showed that measurement precision is limited because the information extraction process must change the datum while measuring it. 20th Century physics has questionably extended that to the attribution of a real uncertainty, not merely one of measurement limitation. Then, the zero of the vacuum would also not be precisely zero but a fluctuation in the Heisenberg uncertainty amount about zero. The Lamb shift was attributed to orbital electron interaction with that fluctuation.

The Lamb Shift, is actually caused by the reduction in the negative potential energy due to the orbital electron being near enough to the nucleus that the full Coulomb effect, as when the incoming wave is plane, is slightly reduced as developed above. There being at small values of r marginally less Coulomb attraction, the energy pit in which the electron resides is less deep, which means that its energy is somewhat more than would otherwise be the case. The amount of the effect decreases with increasing r because the reduction in the Coulomb effect decreases as r increases.

The Lamb Shift occurs at much larger values of r (electron orbit radii that are on the order of $r = 10^{-10} m$) than the quite small value of r at which the neutron forms from the combining proton and electron (on the order of $r = 10^{-15} m$). Nevertheless, the Lamb Shift can be used for an approximate calibration of the above `Reduction Factor`. The Lamb Shift is depicted in Figure 4, below.

The shift is stated in terms of the wave number (reciprocal wavelength) because the Rydberg expression for the spectral lines is in terms of wave numbers. The amount of the *Balmer Â* shift is 0.033 cm^{-1}. That occurs at the $n = 2$ level where the overall level itself has the term value the Rydberg constant divided by n^2. The fractional shift is then as follows.

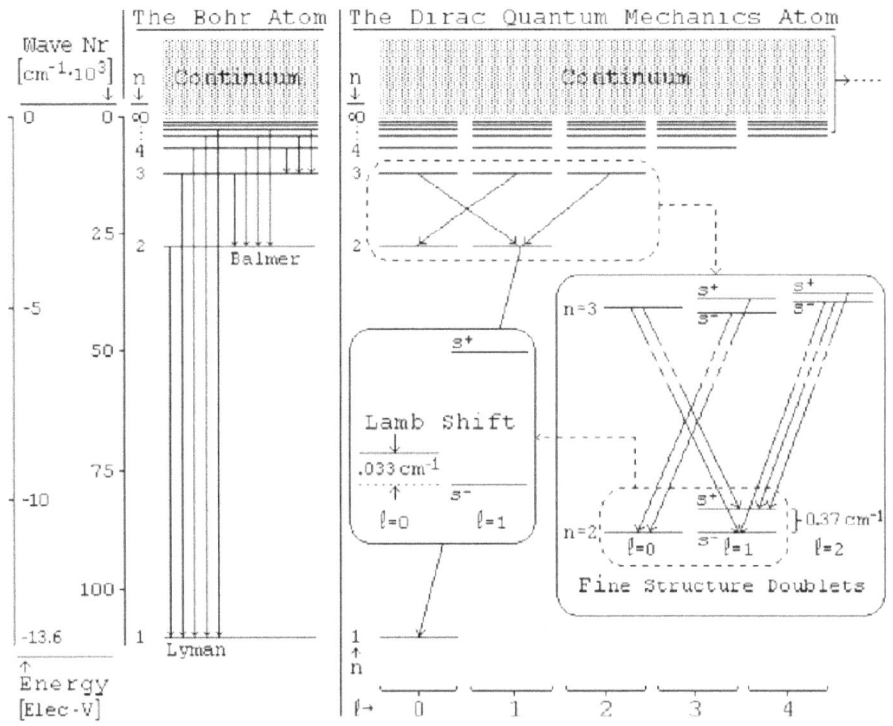

Figure A-1-4
Hydrogen Spectra and the Lamb Shift

$(A-1-21)$ ΔE = Shift = 0.033 cm^{-1} [n=2 Balmer $\hat{A}$ shift]

E = Total Wave Number

$$= \frac{Ry}{n^2} = \frac{109{,}737.31534}{4} = 27{,}434.3 \ cm^{-1}$$

$$\text{Fractional Shift} = \frac{\Delta E}{E} = \frac{0.033}{27{,}434.3}$$

$$= 1.2 \cdot 10^{-6} \text{ [dimensionless ratio]}$$

The above `Fractional Shift` is the fractional energy change to the "normal" Coulomb potential energy due to the effect of the two charges being near to each other. The `Reduction Factor` as used in this analysis, equation `(A-1-15)`, is the net energy after that change, `[1 - the above Fractional Shift]` as follows.

94

$$(A\text{-}1\text{-}22) \quad \left[\begin{array}{c} \text{Reduction} \\ \text{Factor} \end{array} \right] = [1 - \text{Fractional Shift}]$$

$$= 1 - 1.2 \cdot 10^{-6}$$

$$= 0.999,998,80$$

$$\cfrac{1}{1 + \cfrac{1}{\chi \cdot [r - \rho]}} = 0.999,998,80$$

The radius of the $n = 2$ orbit of Hydrogen is $r = 2.1190152 \cdot 10^{-10}$ m. The ρ in the *Reduction Factor* formula is negligible in the case of the Lamb Shift where $r \approx 10^5 \cdot \rho$ and the precision of the Lamb Shift datum is only two significant digits. Equation *(A-1-20)* can then be solved for the value of χ as follows.

$$(A\text{-}1\text{-}23) \quad \chi = \frac{\text{Reduction Factor}}{r \cdot [1 - \text{Reduction Factor}]}$$

$$= 3.9 \cdot 10^{-15}$$

The general formulation for the *Reduction Factor* is, then, the expression of equation *(18)* with the equation *(22)* value of χ substituted and $\rho = r/2$. The expression for the potential energy as the proton and the electron approach each other to form a neutron is then equation *(A-1-15)* with that *Reduction Factor* substituted. That expression can then be solved for r, the $r_{separation}$ with the following result.

$$(A\text{-}1\text{-}24) \quad r_{separation} = 1.3 \cdot 10^{-15} \text{ meters}$$

The precision of this result is limited to the two significant digits of the Lamb Shift datum. Nevertheless, it is quite close to the wavelength of the proton oscillation in the neutron per equation *(A-1-12)*, $\lambda_p = 1.321,408,96 \cdot 10^{-15}$ *meters*.

Alternatively, if $r_{separation}$ is set at λ_p the resulting value for χ can be calculated and from that the value of ΔE, the Lamb Shift. That calculation gives a Lamb Shift of *.033,611,416* cm^{-1} compared to the actual datum of *.033* cm^{-1}.

Two conclusions result from these calculations.

First:
> The cause of the Lamb Shift is the change in the magnitude of the Coulomb effect when the two charges are near to each other not the "radiative coupling of the electron to the zero point fluctuation of the vacuum".

Second:
> The neutron is the combination of a proton and an electron exactly as if each brings to the union its mass equivalent of its escape velocity kinetic energy from the other, the boundary at which the two combine being the wavelength of the proton oscillation, the resulting neutron oscillation being as Figure A-1-1(b) and equation A-1-1 with f_p and f_e being the frequency equivalents of the masses of equations *A-1-10* and *A-1-9* respectively.

95

$\longrightarrow$

Appendix A-2

The Atomic Nuclei

Having now the model of the neutron as a combination of a proton and an electron into a new type of *Spherical-Center-of-Oscillation* oscillating as a resultant of its separate components' *Spherical-Centers-of-Oscillation*, a similar model is now available for the atomic nuclei. In fact, such a model is essential. The "bunch of grapes" and related concepts of the nucleus with its distinct protons and neutrons, simply will not work.

The problems with the "bunch of grapes" concept, whether the "grapes" (protons and neutrons) are closely packed or are a loose assembly in motion relative to each other, are as follows.

1- The *Propagated Outward Flow* (the electric field) of individual protons is partially blocked by the other particles in the nucleus (protons and neutrons). It is not possible to find a configuration of protons and neutrons as distinct entities, collectively constituting an atomic nucleus, where the full net positive electric field of the protons is present in all directions simultaneously.

   Yet, of course, the full field simultaneously in all directions is required, if only for the sake of the orbital electron structure.

2- In the "bunch of grapes" model there is the need for "nuclear binding energy" to overcome the tendency of the nucleus to fly apart because of the mutual repulsion of the protons. The mass deficiency of atomic nuclei is hypothesized as the means of nuclear binding, but an acceptable mechanism is needed.

   To meet that requirement a force, the "strong nuclear force" is further hypothesized, a force that: operates only over the very short distances within a nucleus, is very strong within the nucleus, and is due to "exchange forces", that is the exchange of particles called mesons between the nuclear components. The hypothesis is not entirely convincing, however, and has not been proven.

   Furthermore, as will shortly be seen, there is very little correlation between the amount of mass deficiency and the relative stability of a nucleus.

3- The fact that a free neutron, one not part of an atomic nucleus, decays into a proton and an electron with a modest mean lifetime before decay but that the neutron is entirely stable when a component of a stable atomic nucleus is unexplained and would appear to be unexplainable in the "bunch of grapes" model.

All of those problems are overcome by the *Spherical-Centers-of-Oscillation* nuclear model. The nucleus is a new unitary particle, the resultant of the natural oscillations of its component simple *Spherical-Centers-of-Oscillation* particles [protons and electrons] analogous to the structure of the neutron.

That structure of atomic nuclei, that model, performs as follows.

1- It naturally exhibits the correct electric field in all directions at all times. There are no component particles to get in the way. The nucleus is a single unitary particle with its *Propagated Outward Flow* natural field. It is a single *Spherical-Center-of-Complex-Oscillation*.

2- There is no need for nuclear binding energy, no need for a force to hold component particles together. The nucleus is one (complex) *Spherical-Center-of-Oscillation* not mutually repelling multiple particles.

3- Within the nucleus the neutron does not exist as a separate particle. There is no neutron, as such, within the nucleus at all. Only the neutron components' oscillations are part of the overall atom's nucleus components' oscillations.

There are other advantages to this nuclear model. It is difficult to envision matter-antimatter annihilation of an atomic nucleus and its antimatter counterpart in other nuclear models. How could each particle and its anti-particle get together in a "bunch of grapes" configuration? But the single unitary *Spherical-Center-of-Oscillation* model readily accommodates the mechanism of mutual annihilation presented in Appendix C, *Why No Immediate Mutual Annihilation*.

As will shortly be developed, the *Spherical-Center-of-Oscillation* model accounts for: all of the various nuclei, their masses, their stability or instability, radioactivity and its mean lifetime before decay. It also correlates directly with the origin of the universe. The nuclear model is a complex *Spherical-Center-of-Oscillation* the combination of its co-located component *Spherical-Centers-of-Oscillation*.

Those components are quantity A of protons and quantity $[A-Z]$ of electrons, not the traditional Z protons and $N = [A-Z]$ neutrons. The neutron is itself a combination particle, the combining of one proton and one electron into a new, complex center-of-oscillation per Appendix A-1, *The Neutron*. The fundamental "building block" particles are the proton and the electron. The neutron is more properly viewed as the nucleus of the atom of $Z = 0, A = 1$.

Developing the *Spherical-Center-of-Oscillation* atomic nuclear model in detail proceeds as follows.

THE NUCLEAR SPECIES MODEL

The problem in developing the details of the general nuclear model is: how do multiple protons or multiple electrons, alone, combine into a super *Spherical-Center-of-Oscillation* or as part of one ?

In the case of the neutron the combining of the two component co-located *Spherical-Centers-of-Oscillation*, a proton and an electron, consisted of the direct addition of the two oscillations, the two wave forms. Because the frequencies of the two component wave forms were different the relative phase of the two was not of consequence and the two frequencies *beat* together producing the neutron wave form.

To develop a corresponding general expression for atomic nuclei requires dealing with <u>multiple</u> protons [all of the same frequency] and <u>multiple</u> electrons [all of the same frequency] .

The principal factors determining the form of the model must be the proper representation of *Z* and *A* so that the nuclear electric charge and Coulomb effect are correct *(for Z)* and so that the atomic mass number *(A)* is the nearest integer to the actual exact mass. As developed in Section 3, *The Action of Matter - Coulomb's Law* the *Z* must be the average value of the oscillation. Neither the frequency nor the amplitude of the oscillatory part of the oscillation can affect the value of *Z*.

With regard to *A*, for several reasons one would expect that a "double proton" would have a frequency of twice the normal single proton's frequency. A double proton would be expected to have approximately twice the mass of a single proton. Since it has already been found that mass is proportional to frequency the double mass would seem to call for a doubling of the frequency.

One would then expect that a double proton would have the wave form of a single proton except that its average value would be double (its *Z* would be *Z* = *2*), and its oscillation frequency would be doubled. In general by this reasoning, a particle that is *M* multiples of a fundamental particle such as a proton or an electron would have *M* times the average value and *M* times the frequency of the basic particle as in equation *A-2-1*, below.

$$(A-2-1) \qquad U\big[M \text{ protons}\big] = U_c \cdot \Big[M - Cos(2\pi \cdot [M \cdot f_p] \cdot t)\Big]$$

$$U\big[M \text{ electrons}\big] = -U_c \cdot \big[M - Cos(2\pi \cdot [M \cdot f_e] \cdot t)\big]$$

Then, the structure of an atomic nucleus would be

$$(A-2-2) \qquad U\big[{}_Z Sym^A\big] = A \text{ protons} + [N = A - Z] \text{ electrons}$$

$$= U_c \cdot \Big[A - Cos(2\pi \cdot [A \cdot f_p] \cdot t)\Big] + \Big[-U_c \cdot \big[N - Cos(2\pi \cdot [N \cdot f_e] \cdot t)\big]\Big]$$

$$= U_c \cdot \Big[Z - Cos(2\pi \cdot A \cdot f_p \cdot t) + Cos(2\pi \cdot N \cdot f_e \cdot t)\Big]$$

where the "Sym" of ${}_Z Sym^A$ means the element *symbol*, the one or two letter abbreviation for the element name.

With regard to the nuclear species formulation of the above equation *A-2-2*:

1- The formulation reduces to the form for a neutron per equation *A-1-1* when the parameter values are *A = 1, Z = 0*.

2- The formulation yields the proper overall average value of the wave form, *Z·U_c*, which corresponds to the net positive charge of the nuclear *Spherical-Center-of-Oscillation*.

3- As is necessary for the electric charge *[Z]* of the nucleus to be independent of the mass *[A]* of the nucleus the amplitude of the oscillatory portion of the expression (the amplitude of each of the two oscillatory terms of equation *A-2-2*) is the same for all nuclear species and not a function of *A* or *Z*.

The resulting conclusion from this is:

The amplitude of the Spherical-Center-of-Oscillation, U_c, is a universal constant the same constant quantity that is the cause of the fundamental electric charge, q, being a constant. (U_c and q are essentially the same thing.)

Figures A-2-2 through A-2-4 on the following pages depict the wave forms per equation $A-2-2$ of the principal isotopes of the first three elements, $z = 1$ to 3, Hydrogen, Helium, and Lithium. The neutron, being in effect the element $z = 0$ is depicted to the same scale in Figure A-2-1, below. The electron oscillation is included in that figure for comparison purposes.

The graphs use a ratio of f_p/f_e of $9/1$ rather than the much larger actual value, which is on the order of the rest value, $1,836.152,701$. The $9/1$ ratio permits indicating the general variation of the wave form in a moderate amount of space. At the actual f_p/f_e ratio the wave form change from f_p cycle to f_p cycle is much more gradual than in the figures.

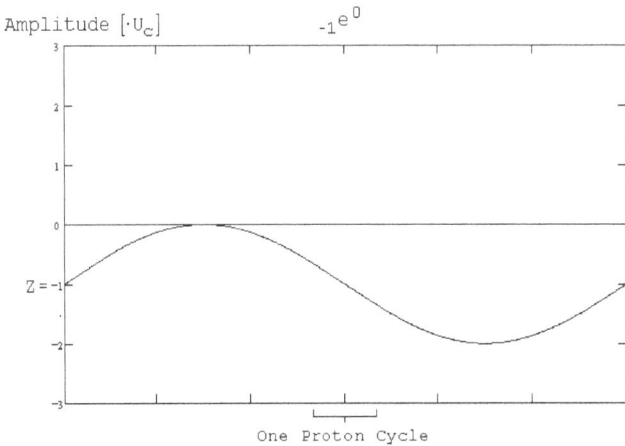

Figure A-2-1(a), The Electron Wave Form

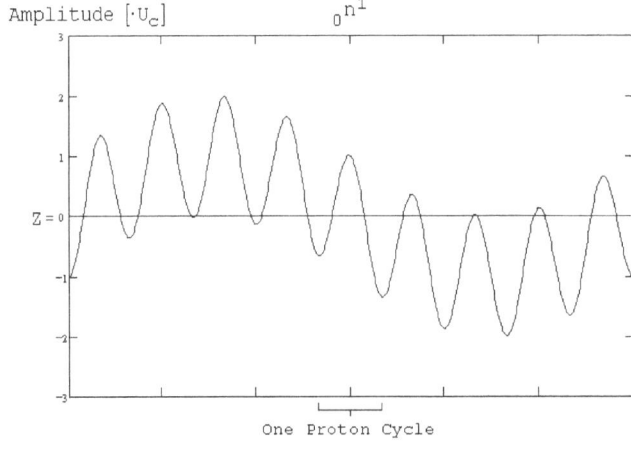

Figure A-2-1(b), The Neutron Wave Form

100

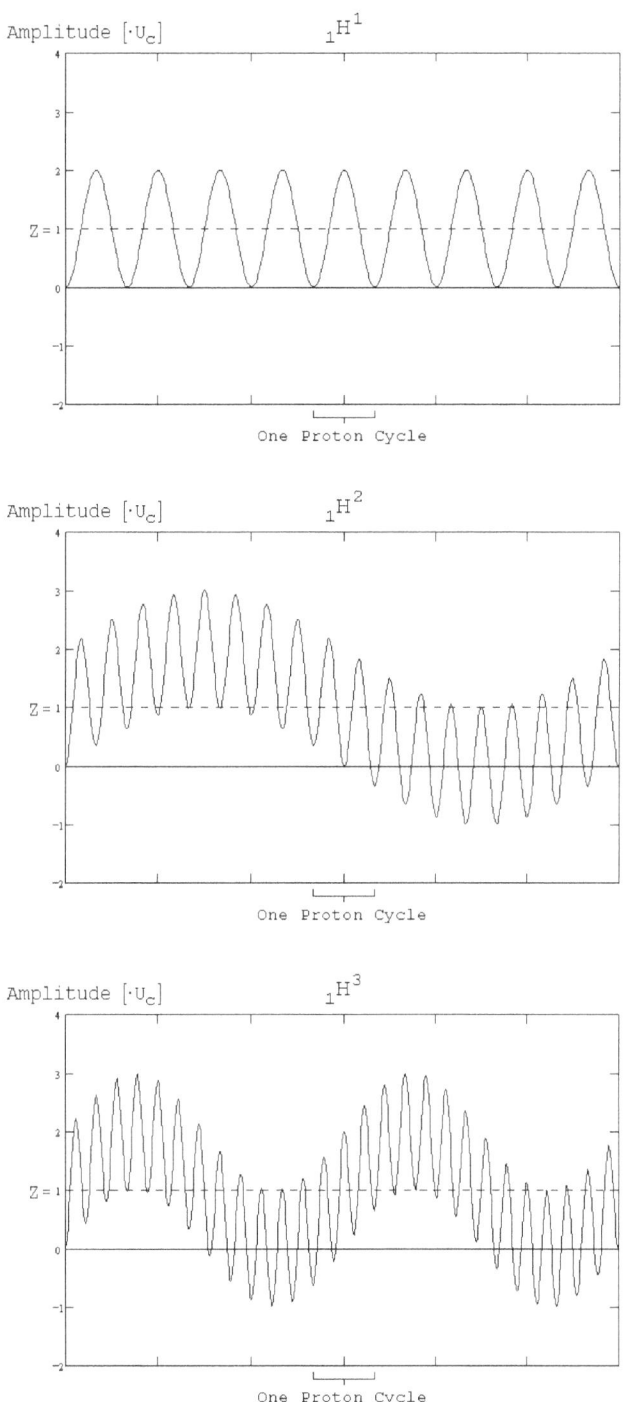

Figure A-2-2, The Hydrogen Wave Forms

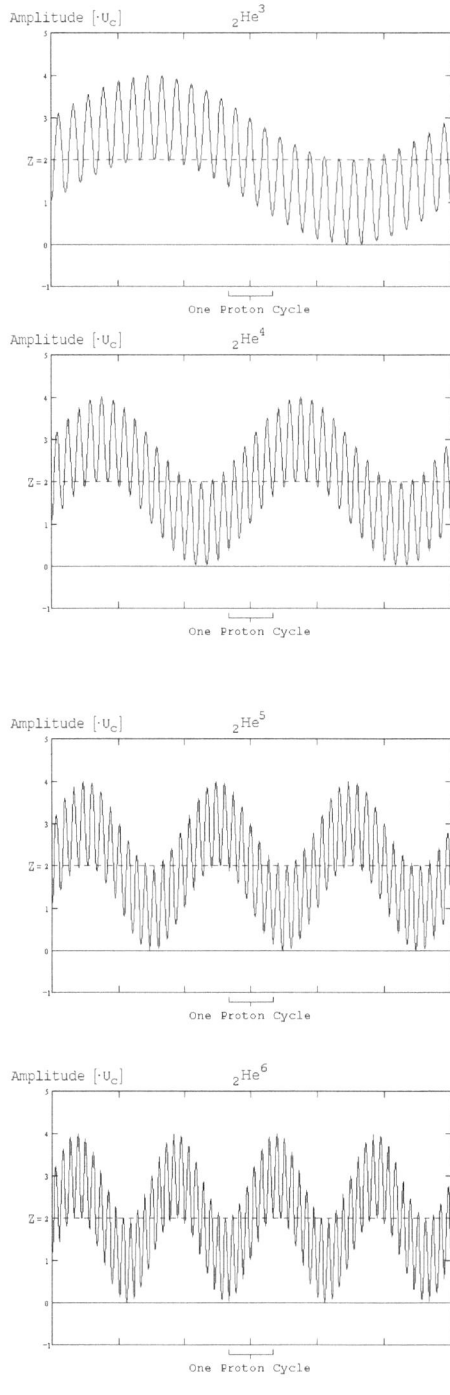

Figure A-2-3, The Helium Wave Forms

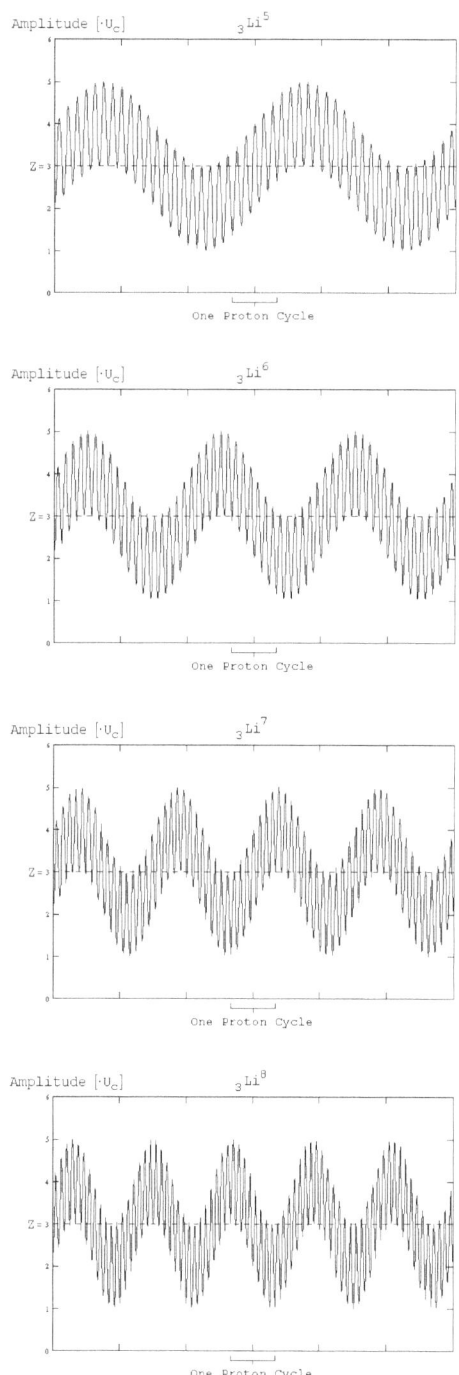

Figure A-2-4, The Lithium Wave Forms

103

FURTHER DEVELOPMENT OF THE NUCLEAR SPECIES MODEL

In extrapolating the model of the neutron developed in the preceding Appendix A-1 to the family of all the nuclear species, the mathematical description of the neutron model has been treated, but not the remainder, the formation of the neutron from its component proton and electron mutually accelerating toward each other, uniting into a new *Spherical-Center-of-Oscillation.*

Of course, unlike the case of the neutron, the components of an atomic nucleus, protons and electrons, cannot naturally and unaided come together to form a nucleus. The description of these atomic nuclei in terms of their component protons and electrons assembling in a particular manner is not to say that that action actually occurs in that way. Rather, it is a procedure for determining what are the characteristics of the resulting nucleus.

There are only two ways that such a nucleus can come into existence. One is through the process of radioactive decay of a more complex nucleus which was the case with the "Big Bang".

In Section 2 under the sub-title "The Form of Matter as Generated by the Big Bang" the Big Bang is described as follows:

"Judging by its result, the "Cosmic Egg" was not unlike an immense atom, a very unstable immense atom [as are all of the atomic species of atomic number exceeding 83 which the cosmic egg would have immensely exceeded]. Its "Big Bang" was a kind of explosive nuclear decay. Such decays follow chains:

- From a heavy and complex composition,

- Through many various stages of multiple less heavy less complex products,

. . . . .

- Until ultimately they arrive at many multiple stable forms [and some long half-life still slowly decaying forms].

Some of those decay chains ended in stable species heavier than Hydrogen, heavier than the proton. Those appear to us as the various stable atomic species of the Periodic Table of the Elements."

The other way for some of the complex atomic nuclei to form is for the set of components, or more likely some two less complex nuclei, to be accelerated toward each other with so much energy that they merge in spite of their mutual repulsion. This is thought to occur in stars where the product is later spread to the universe by the star exploding as a super nova.

As in the case of the neutron, escape velocity masses are a factor in all of the atomic nuclei. The point when two *opposite* charged particles mutually attracting each other, *i.e.* the case of the neutron, achieve their mutual escape velocity is just before they collide. But, when two *like* charged particles such as two protons, rushing away from each other in mutual repulsion, achieve their escape velocity is at their maximum separation (infinite distance). For them to stay together and to not so rush apart, they must lose their mutual escape velocity kinetic energies.

Evaluation of those cases in the manner as was done for the neutron becomes an inordinately complex problem. The escape velocity calculations, which involve relativistically calculating the potential energy relationships among the particles, and the resulting velocities that they take on become quite complex when more than two particles are involved.

Deuterium illustrates the simplest case of multi-body escape velocity difficulty. After the neutron, $_0n^1$, and the Hydrogen nucleus $_1H^1$, the next most complex nucleus is that of the Hydrogen isotope, Deuterium, $_1H^2$, the nucleus of the Deuterium atom, which is also referred to as the deuteron.

The following is not a description of how a deuteron comes to be. Rather it is an analysis of the considerations that must be satisfied for it to exist regardless of how it came to be.

The deuteron consists of the combination of two protons and one electron. Those two protons mutually repel each other. Between the electron and each of the protons there is attraction. Figure A-2-5, below, illustrates the Deuterium nucleus component particles' configuration for their approach to merger into a deuteron. Their mutual repulsion places the two protons on opposite sides of the electron which is in the center by default, where it equally attracts each of the protons.

Legend:

>——< Proton — Electron Coulomb Attraction
        (as in the case of a simple neutron)

←——→ Proton — Proton Coulomb Repulsion

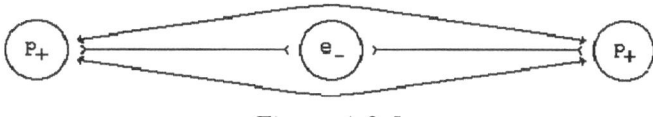

Figure A-2-5

If the mutual repulsion between the two protons is ignored for the moment, then the same escape velocity kinetic energy as was found in the case of the formation of a neutron would be developed between each of the protons and the single electron. The velocity situation will be different from that in the neutron case because the electron, being simultaneously and equally attracted in exactly opposite directions experiences no net acceleration. But the energies are the same, all of it appearing in kinetic energy of the protons.

Thus for this part of the interaction the deuteron should have a mass excess equal to twice the mass excess of the neutron,

(A-2-3) $2 \cdot [839.854 \; \mu\text{-}amu]$ or $0.001,679,708 \; amu$.

The Deuterium nucleus' mass deviation from the mass of its component particles is calculated in the same way as was done for the neutron at equation *D-2*, as follows.

(A-2-4) $\Delta m_{De} = \left[m_{De,atom} - m_{De,orbital \; electron} \right]_{rest} - \left[2 \cdot m_p + m_e \right]_{rest}$

$= 2.014,101,779 - 0.000,548,579,903 \dots$

$\dots -2 \cdot [1.007,276,470 + 0.000,548,579,903]$

$= -0.001,548,319 \; amu$

105

This negative numerical result is to be expected. Unlike the neutron, the deuteron must have a mass deficiency, which is always found when the nucleus contains multiple particles needing to be held together. In this Deuterium case the mass of the atomic nucleus has a net mass decrease in the amount per equation *A-2-5*, below, which can only be due to the interaction of the two protons.

```
(A-2-5)   +1,679.708 µ-amu proton-electron mass increase Equation A-2-3
        -  -1,548.319 µ-amu deuteron net mass decrease of Equation A-2-4
          ────────────
          +3,228.027 µ-amu required proton-proton mass decrease
```

Theoretical Formation of a Deuteron From Its Components

The combination of the proton - electron Coulomb attraction and the proton - proton Coulomb repulsion produces the resulting deuteron net mass. That decrease is equal to the energy that the two component protons lose to become part of the nucleus as partially offset by the mass excess developed between the electron and each of the protons.

However, there is much more to the overall development of this nuclear model. That further development requires a detailed analysis of the various nuclear types and a matching of the nuclear model to that data.

NUCLEAR DATA ANALYSIS

Table A-2-6, starting on the second following page, is a partial compilation of all atoms known, stable and unstable. (The table is a partial list plus a reference to the source data for a complete compilation.) In the table the atoms are grouped by atomic number, Z, with Z listed in ascending order and with all atoms having the same Z listed in ascending order of their atomic mass number, A.

Just as for the fundamental physical constants, whose values are published by CODATA, the best set of internationally accepted masses of the various atoms are up-dated and published as more accurate measurements become available. The present data is from *The 1983 Atomic Mass Evaluation* by The National Institute of Nuclear Physics and High-Energy Physics, Amsterdam; University of Technology, Delft, The Netherlands; and Laboratoire Rene Bernas du CSNSM, Orsay, France. The atomic masses of Table A-2-6 are those of that 1983 evaluation. All of the data are rest masses.

In addition to listing for each atom its symbol, name, Z, A, and mass, the table also indicates whether the atom is stable or not, the particle emitted if the atom is unstable, and the *mass deficiency* and the *separation energy*, defined in equations *A-2-6* and *A-2-7*, below.

```
(A-2-6)                    Mass Deficiency

    Mass         Mass of        Mass of      Atomic
    Defic-   =   Nuclear    +   Orbital   -    Mass
    iency        Components     Electrons   (Per Table)

   [Here for mass deficiency the "Nuclear Components" are
      Protons and neutrons.]
```

The *separation energy* deals with the possibility of decay of a nuclear type. It is the mass of the nucleus before decay less the mass of the decay products as equation A-2-7, below.

(A-2-7)   Separation Energy
      Separation = Mass of Nucleus Before
        Energy   Decay

                + One electron mass if the decay
                  is by the nucleus capturing an
                  electron

                - Mass of resulting nucleus
                - Mass of particle(s) emitted

    [The nuclear mass is in each case the
    atomic mass per the table less the mass of
    the Z orbital electrons.]

If the *separation energy* is positive then the initial component(s) have enough mass to make up the final components plus some extra mass to appear as energy of motion of the final components or as a particle or wave radiation. If the *separation energy* is negative then the decay cannot take place because there is not enough mass to make up the final components and conservation would be violated.

Therefore, positive *separation energy* means instability and negative *separation energy* means stability.

In a practical sense the positive *separation energy* must be large enough to supply the escape velocity of the product particles, just as was the case for the neutron. Otherwise a decay would be followed by an immediate recombination and be, in effect, no decay at all.

Any nuclear species except Hydrogen and the neutron can have, at least in theory, a family of separation energies for different decay products. The separation energy listed in Table A-2-6 is the largest one for that atom, which corresponds in general to the most probable decay, which is the decay listed in the table.

Examination of the *mass deficiency* data in Table A-2-6 discloses insufficient correlation with the various atoms' nuclear stability or instability. Mass deficiency tends generally to increase with atomic number, Z, and atomic mass number, A, but there is no value of mass deficiency that separates stable and unstable nuclei.

It would probably be more appropriate to work in terms of the mass deficiency per nuclear particle *[MD/A]* or per nuclear proton *[MD/Z]* since it would presumably require more binding energy to bind more protons while neutrons (neutralized protons) are not so much part of the problem. But neither of those values show sufficient correlation to specifically relate them to nuclear stability or instability.

Nuclear stability / instability correlates with mass deficiency in only a broad and general sense.

The data in Table A-2-6 make clear that, without yet asking for a reason (which is presented below), *separation energy* is the touchstone of nuclear stability. For each Z there is a number of nuclear species, isotopes, of successively larger A. They differ among each other only by the number of neutrons in the nucleus. The number of protons is the same for the same Z. For any Z the isotopes of "medium values of A" are stable. They have negative *separation energy*; that is, the total mass / energy of the nucleus is not large enough to make up any set of decay products whatsoever.

107

Table A-2-6
The Natural Atomic Species and Masses

		Z	A	Measured Atomic Mass amu	Emission if any	Mass Defic'y μ-amu	Separ'n Energy μ-amu
n	Neutron	0	1	1.008,664,904	-Beta	-840	840
H	Hydrogen	1	1	1.007,825,035		0	(-)
			2	2.014,101,779		3,228	(-)
			3	3.016,049,27	-Beta	9,106	20
He	Helium	2	3	3.016,029,31		9,965	(-)
			4	4.002,603,24		32,056	(-)
			5	5.012,220	Neutron	29,425	952
			6	6.018,886,0	-Beta	31,424	3,765
Li	Lithium	3	5	5.012,540	Proton	28,265	3,209
			6	6.015,121,4		34,348	(-)
			7	7.016,003		42,132	(-)
			8	8.022,485,6	-Beta	44,314	17,180
			9	9.026,789,0	-Beta	48,676	14,607
Be	Beryllium	4	6	6.019,725	Proton	28,905	457
			7	7.016,928,3	Elec Capt	40,367	2,022
			8	8.005,305,12	Alpha	60,655	2,842
			9	9.012,182,2		62,443	(-)
			10	10.013,534,1	-Beta	69,756	597
			11	11.021,658	-Beta	70,297	12,353
B	Boron	5	8	8.024,605,8	+Beta	40,514	19,301
			9	9.013,328,8	Proton	60,456	1,296
			10	10.012,936,9		69,513	(-)
			11	11.009,305,4		81,809	(-)
			12	12.014,352,6	-Beta	85,427	14,353
			13	13.017,802	-Beta	90,642	14,447
C	Carbon	6	10	10.016,856,4	+Beta	64,754	3,322
			11	11.011,433,3	+Beta	78,842	2,128
			12	12.000,000,000		98,940	(-)
			13	13.003,354,826		104,250	(-)
			14	14.003,241,982	-Beta	113,028	168
			15	15.010,599,2	-Beta	114,335	10,490
			16	16.014,701	-Beta	118,898	9,601
N	Nitrogen	7	12	12.018,613,0	+Beta	79,487	18,613
			13	13.005,738,60	+Beta	101,026	2,384
			14	14.003,074,002		112,356	(-)
			15	15.000,108,97		123,986	(-)
			16	16.005,099,9	-Beta	127,660	10,185
			17	17.008,450	-Beta	132,974	9,319

Table A-2-6 (continued)

		Z	A	Measured Atomic Mass amu	Emission if any	Mass Defic'y μ-amu	Separ'n Energy μ-amu
O	Oxygen	8	14	14.008,595,33	+Beta	105,994	5,521
			15	15.003,065,4	+Beta	120,189	2,956
			16	15.994,914,63		143,724	(-)
			17	16.999,131,2		148,172	(-)
			18	17.999,160,3		156,808	(-)
			19	19.003,577	-Beta	154,337	5,174
			20	20.004,075,5	-Beta	162,504	4,094
F	Fluorine	9	16	16.011,466	+Beta	119,614	16,551
			17	17.002,095,05	+Beta	137,650	2,964
			18	18.000,937,4	+Beta	147,472	1,777
			19	18.998,403,22		166,230	(-)
			20	19.999,981,39	-Beta	165,758	7,546
			21	20.999,948	-Beta	174,456	6,105
Ne	Neon	10	18	18.005,710	+Beta	141,860	4,773
			19	19.001,879,7	+Beta	154,355	3,476
			20	19.992,435,6		180,862	(-)
			21	20.993,842,8		188,120	(-)
			22	21.991,383,1		199,245	(-)
			23	22.994,465,4	-Beta	196,429	4,698
			24	23.993,613	-Beta	205,946	2,652
Na	Sodium	11	20	20.007,344	+Beta	156,716	14,908
			21	20.997,650,5	+Beta	175,074	3,808
			22	21.994,434,1	+Beta	186,955	3,051
			23	22.989,767,7		209,525	(-)
			24	23.990,961,4	-Beta	207,758	5,919
			25	24.989,953	-Beta	217,431	4,216
			26	25.992,586	-Beta	223,463	9,992
Mg	Magnes-sium	12	22	21.999,574,3	Proton	180,975	5,689
			23	22.994,124,4	+Beta	195,090	4,357
			24	23.985,042,3		222,915	(-)
			25	24.985,737,4		230,885	(-)
			26	25.982,593,7		242,6	(-)
			27	26.984,341,2	-Beta	239,533	2,803
			28	27.983,876,8	-Beta	248,662	1,967
Al	Aluminum	13	24	23.999,941	+Beta	197,099	14,899
			25	24.990,429,0	+Beta	215,275	4,692
			26	25.986,892,2	+Beta	227,477	4,299
			27	26.981,538,6		252,414	(-)
			28	27.981,910,2	-Beta	249,789	4,983
			29	28.980,446	-Beta	259,918	3,951
			30	29.982,940	-Beta	266.089	9.170

109

Table A-2-6 (continued)

	Z	A	Measured Atomic Mass amu	Emission if any	Mass Defic'y μ-amu	Separ'n Energy μ-amu
Si Silicon	14	26	25.992,330,00	+Beta	221,200	5,438
		27	26.986,703,90	+Beta	235,491	5,165
		28	27.976,927,10		253,932	(-)
		29	28.976,494,90		274,787	(-)
		30	29.973,770,10		274,419	(-)
		31	30.975,362,10	-Beta	281,492	1,600
		32	31.974,148,30	-Beta	291,371	241
P Phospho-	15	28	27.992,313,00	+Beta	237,707	15,386
rus		29	28.981,803,00	+Beta	256,881	5,308
		30	29.978,306,70	+Beta	269,043	4,537
		31	30.973,762,00		294,850	(-)
		32	31.973,906,80	-Beta	290,772	1,836
		33	32.971,725,20	-Beta	301,619	267
		34	33.973,636,20	-Beta	308,373	5,770
S Sulfur	16	30	29.984,903,00	+Beta	261,606	6,596
		31	30.979,554,30	+Beta	275,620	5,792
		32	31.972,070,70		291,769	(-)
		33	32.971,458,43		314,483	(-)
		34	33.967,866,65		313,302	(-)
		35	34.969,031,83	-Beta	320,802	179
		36	35.967,080,62		331,418	(-)
		37	36.971,125,54	-Beta	336,038	5,223
		38	37.971,162,00	-Beta	344,667	3,151

(etc.)   1 μ-amu = 0.000,001 amu
                                   read as "micro-amu"

This table begins a list of the known atomic species. By "known" is meant that a sufficient amount of the isotope has been isolated to enable a measurement of its atomic mass with some reasonable accuracy. The omitted isotopes follow the same patterns as those included.

The data is derived from:

"The 1983 Atomic Mass Evaluation" by The National Institute of Nuclear Physics and High-Energy Physics, Amsterdam; University of Technology, Delft, The Netherlands; and Laboratoire Rene Bernas du CSNSM, Orsay, France.

The table can be extended to its finish by applying the equations *E-6* and *E-7* definitions of separation energy and mass deficiency to the data in "The 1983 Atomic Mass Evaluation".

Those nuclear species of smaller A have positive *separation energy* and emit a particle which in most cases (*+Beta*, a positron) changes the species to being species *[Z-1]* at the same A, a step toward being a species "of medium A" for the new, lower z that it has become. (In some cases a different particle is emitted but the tendency to change toward a species where the A is "medium" is always the case.) For example, unstable species $_7N^{12}$ emits a *+Beta* and becomes stable species $_6C^{12}$.

Likewise, the species of relatively large A for their z also have positive *separation energy*. They in most cases emit a particle (*-Beta*, an electron) which changes the species to being species *[Z+1]* at the same A, a step toward being a species "of medium A" for the new higher z that it has become. For example, unstable species $_7N^{17}$ emits a *-Beta* and becomes stable species $_8O^{17}$.

The particle emission process, *radioactivity*, does not always immediately occur. Rather it is delayed in various amounts and occurs at an overall exponential rate. It is treated in Appendix A-3, *Radioactivity*.

So to speak, all atomic nuclei are unstable; however, there are no products to which those with negative *separation energy* can decay; they are forced into stability by the requirements of conservation of mass / energy. Those with positive *separation energy* can and do decay and the process, the nature of the particle emitted, is such as to move them toward being stable species.

THE STABLE AND UNSTABLE RANGES OF NUCLEAR SPECIES

As indicated greatly exaggerated in Figure A-2-7, below, it is the way that the mass varies from isotope to isotope that results in a narrow range of nuclei with negative *separation energy* and consequent stability, the nuclei on either side of that range having positive *separation energy* and consequent instability.

For a given z the masses of the isotopes are not exactly some constant number times A; rather they vary from such a straight line relationship, only very slightly, in an *"S-shaped"* curve fashion. This curvature in the variation of mass, which is so important and significant, is too small to be observed in a practical unexaggerated plot. If, instead, the plot is of *[A - Exact Nuclear Mass (amu)]* versus *[A]* as in Figure A-2-8, next page, then only the deviations from linearity are plotted, the changes in curvature which range from small to large to small again.

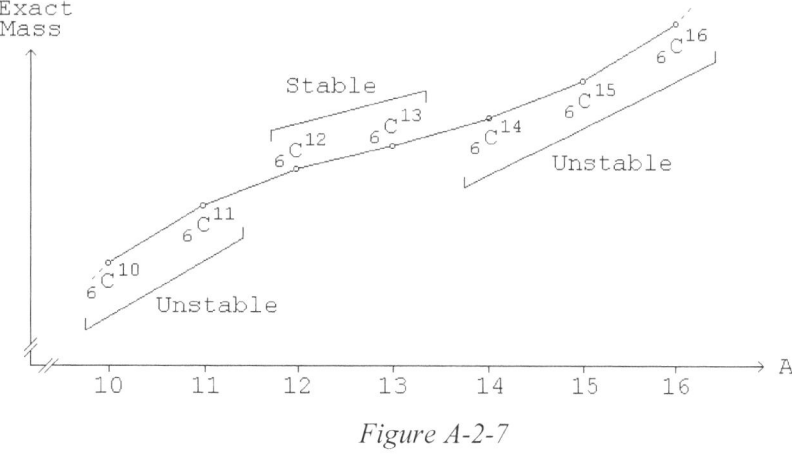

Figure A-2-7

"S-shaped" Curvature in Isotope Nuclear Mass Variations (Exaggerated)

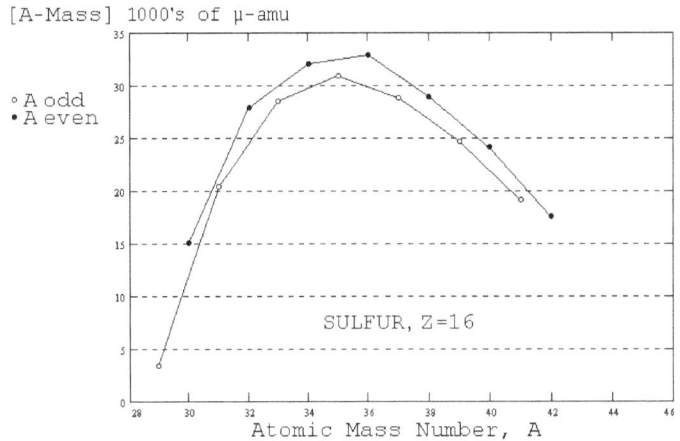

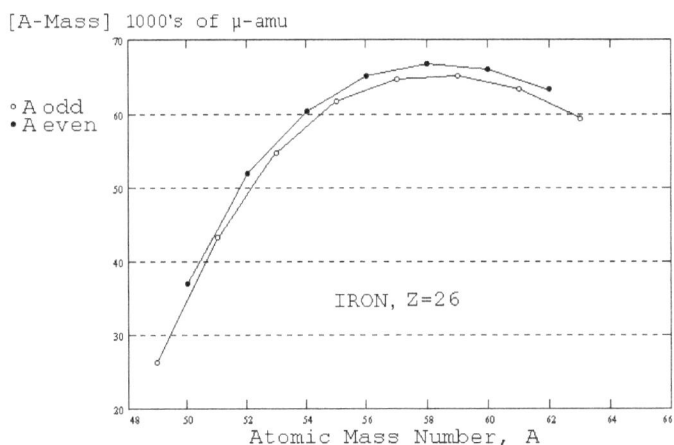

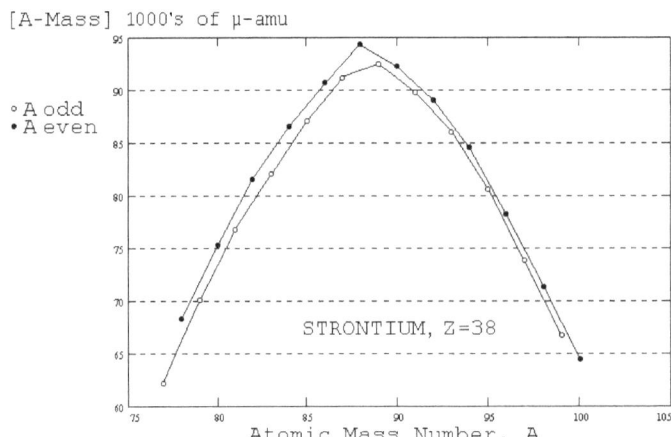

Figure A-2-8, Page 1

[A-Mass] 1000's of μ-amu

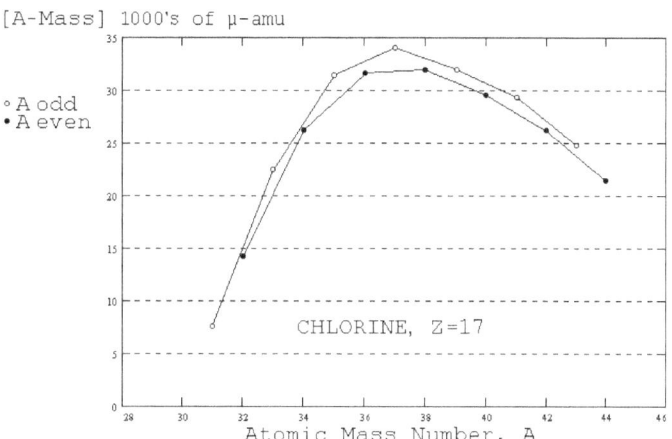

[A-Mass] 1000's of μ-amu

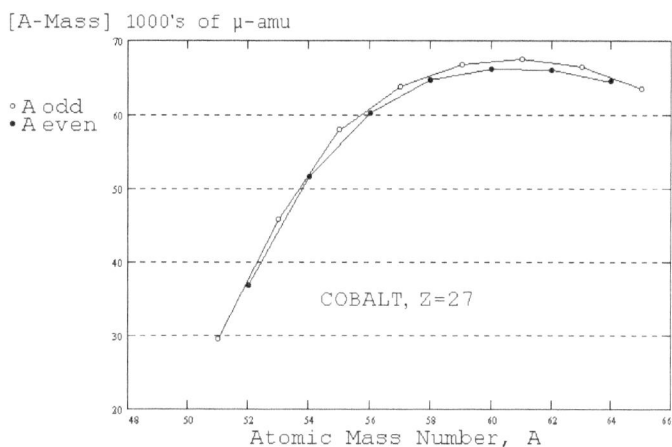

[A-Mass] 1000's of μ-amu

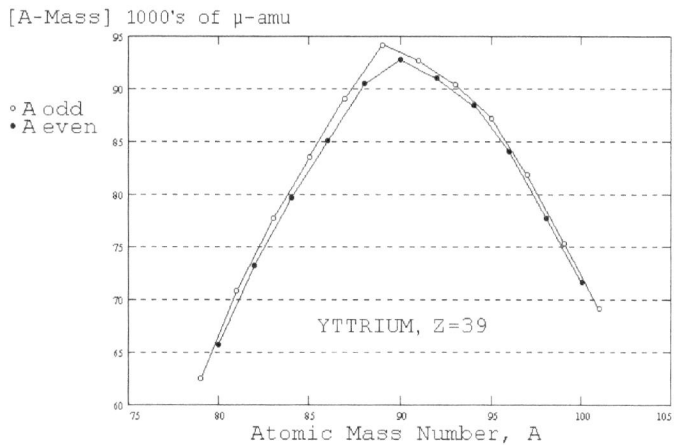

Figure A-2-8, Page 2

These variations in the nuclear species masses are relatively small. Broadly speaking the masses are all very near to A, which varies linearly. Yet these small mass variations account for the entire family of stable isotopes that give us our world and its characteristics. Clearly it is of crucial importance for a model of nature to model and account for the exact nuclear species masses.

THE "SERIES" PATTERN IN THE NUCLEAR SPECIES

The data of Table A-2-6, the masses and the mass deficiencies, appear to be random and chaotic in their minor variations, the very variations that are crucial to accounting for the behavior of matter. But, since nature is orderly, there must be an underlying pattern or patterns that account for the exact actual masses, which are themselves the cause of the overall pattern of stable and unstable species. It is those patterns that must be found and modeled. Their presence is confirmed by the regularity of the curves of Figure A-2-8.

Traditionally N is the number of neutrons in the nucleus, but in terms of *Spherical-Centers-of-Oscillation* and the nuclear model of Equation $A-2-2$, N is the number of electrons entering into the formation of the nucleus [in effect forming neutrons with some of the protons]. If a plot is made to show the trend of $N/_A$ versus A a pattern emerges in the entire family of nuclear species as shown in Figure A-2-9, below.

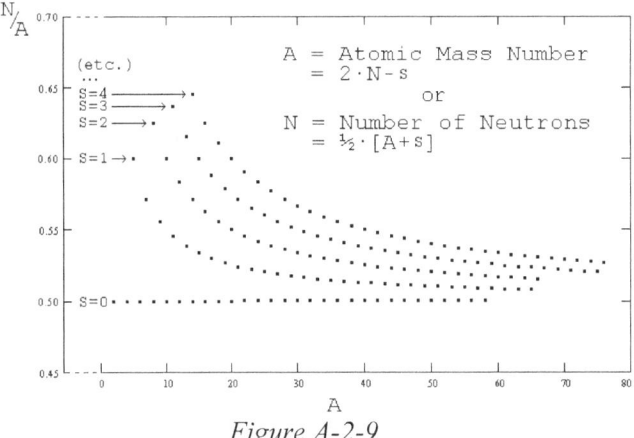

Figure A-2-9

The nuclei appear in *series* according to the relative amounts of the particles, more precisely according to

$(A-2-8)$ $A = 2 \cdot N - s$

where s is an index, a series number for each of the series. The seemingly fairly random pattern of the atomic nuclear species now becomes orderly based only on the ratio of the neutron number to the atomic mass number, $N/_A$. This suggests an underlying structural pattern to the assembly of the various atomic nuclei.

However, Figure A-2-9 is in terms of the integers, A and N, not exact masses. That a set of integers produces an orderly pattern does not necessarily mean that the actual exact masses are orderly. It is a systematic pattern of nuclear structure and exact nuclear masses that must be found. The curves of Figure A-2-8 show such a pattern to exist within families of nuclei of the same Z, but one must exist for all of the nuclei collectively.

Turning now to a comparative examination of the masses of various nuclei within an *s-series* per equation *A-2-8*, analysis is difficult because the search is for patterns of behavior in very minor variations in relatively large quantities. If the overall masses are analytically compared the relatively large total masses prevent observation of the minor mass variations. A procedure to get around that problem is to find a directly related smaller number to analyze.

Such a procedure was used in Figures A-2-8, the related quantity being *[A - mass]*. That same quantity will now be employed again except slightly modified. For Figure A-2-8 the range of values of the masses depicted in one graph was quite limited. Now a much greater mass range is to be treated. In order to reduce the size of graph required for a given precision or resolution in the graph, the related quantity plotted will now be *[A - mass]*/$_A$. Because it has already been found that there is a significant distinction between odd and even *Z* species the two will be analyzed separately. The resulting analysis of selected typical *s-series* of nuclear species is presented in Figure A-2-10, below.

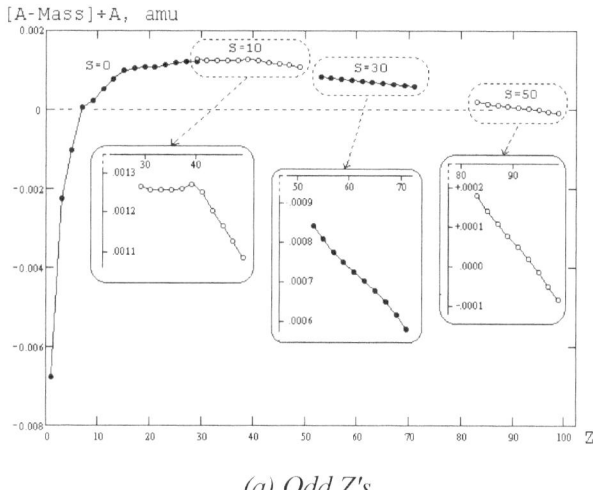

(a) Odd Z's
Figure A-2-10

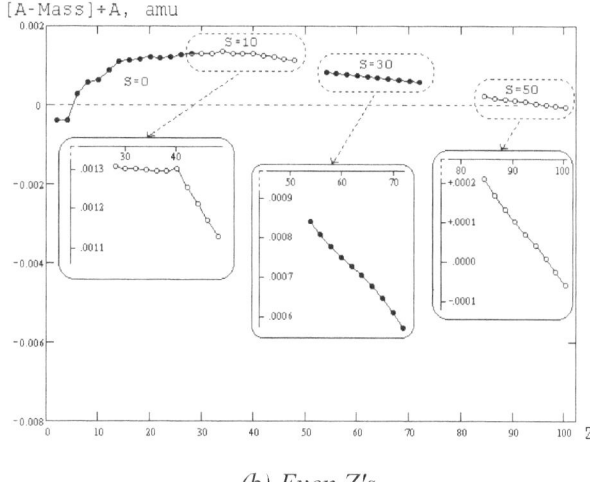

(b) Even Z's
Figure A-2-10

The middle *s series* show the "*S-shaped*" curve of Figure A-2-7. It is present less pronounced in the high *s series* because the changes from species to next higher species are fractionally smaller, that is for large A and large N the ratio N/A is very little different from the ratio $[N+1]/[A+1]$. It appears interrupted at low *s series*.

> These data indicate that there is a simple and regular mode of behavior, structure or process that operates effectively for middle and high Z or high *s series*, that the variations from nuclear type to type are smooth and regular there. That mode appears to also operate for low Z, low *s series*, but is apparently there partially overwhelmed by some other effect not so far detected and taken into account.

THE LOW Z LOW S EFFECT - POLYTOPES

To analyze the process operating at low Z or on low *s series*, <u>Figure A-2-12 on the next two pages</u> investigates the same changes as did Figure A-2-10, but now for several adjacent low *s series*: *s* = -1, 0, and +1. The outstanding characteristics of these data, as plotted, is that regular dips or valleys in the graphs occur at values of Z just following each of **z = 4, 8, and 20**. <u>See Figure A-2-12 then return here.</u>

To understand the effect operating here a brief digression into a relatively slightly attended area of mathematics is necessary. The subject area is that of *polytopes*. A polytope is a geometric figure in *[n]* dimensions having as its boundary a number of geometric figures in *[n - 1]* dimensions. If the boundary figures are all identical then the polytope is *regular*, and it is regular polytopes that are of interest here.

A three - dimensional polytope is a *polyhedron*. Its boundary is flat *faces* that are polygons. Some common polyhedrons are the pyramid and the cube. It turns out that the regular polyhedrons are central to atomic nuclear structure. There are only five regular polyhedrons that can exist, listed in Table A-2-11 below. (They are sometimes referred to as the *Platonic Solids* because Plato was the first to recognize and study them).

Name	Face	Nr Of Faces	Surface Area	Volume	Radius of Inscribed Sphere
Tetrahedron	Equilateral Triangle	4**	$1.73 \cdot a^2$	$0.12 \cdot a^3$	$0.20 \cdot a$
Cube	Square	6	$6.00 \cdot a^2$	$1.00 \cdot a^3$	$0.50 \cdot a$
Octahedron	Equilateral Triangle	8**	$3.46 \cdot a^2$	$0.47 \cdot a^3$	$0.41 \cdot a$
Dodecahedron	Regular Pentagon	12	$20.65 \cdot a^2$	$7.66 \cdot a^3$	$1.11 \cdot a$
Icosahedron	Equilateral Triangle	20**	$8.66 \cdot a^2$	$2.18 \cdot a^3$	$0.75 \cdot a$

Table A-2-11

The Regular Polyhedrons

The appearance in the above table of the same three key numbers: **4, 8, and 20**, that turn up in the graphs of Figure A-2-12 is immediately noticeable. Furthermore, the polyhedrons at which those numbers appear are the three regular polyhedrons that have as face the equilateral triangle, the most simple regular polygon. But, of most significance is that those three cases have relatively the smallest overall sizes, are the most compact.

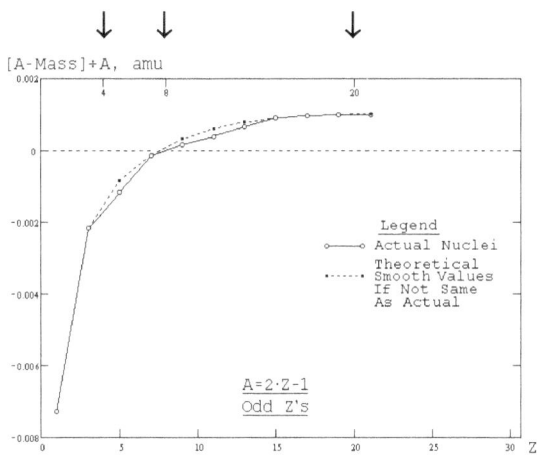

(a) Series s = -1

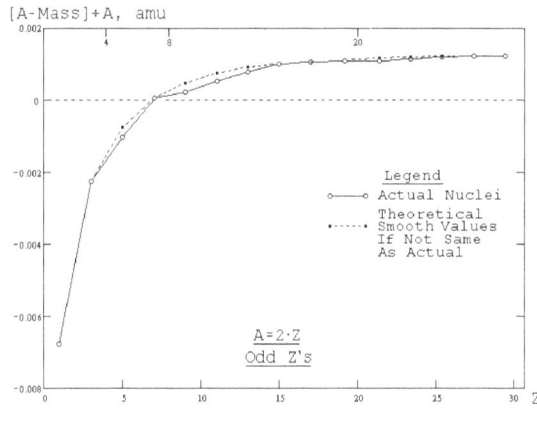

(b) Series s = 0

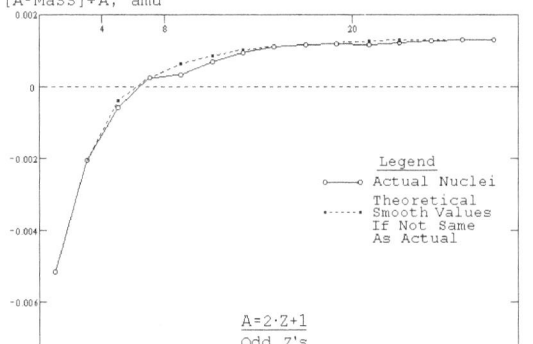

(c) Series s = +1
Figure A-2-12 Odd

117

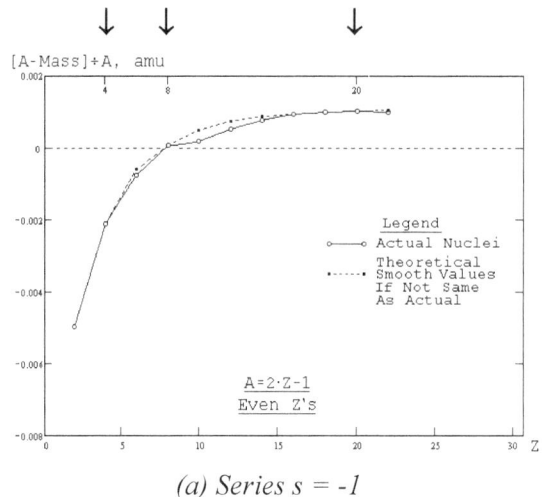

(a) Series s = -1

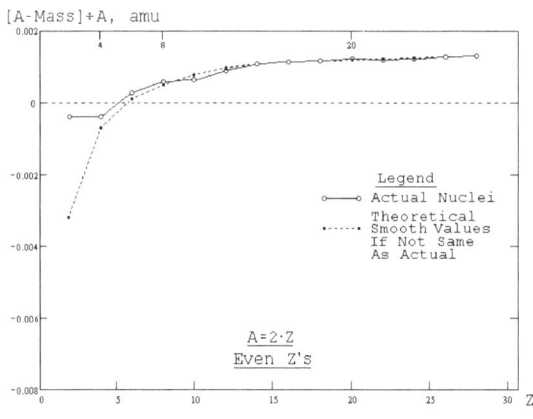

(b) Series s = 0

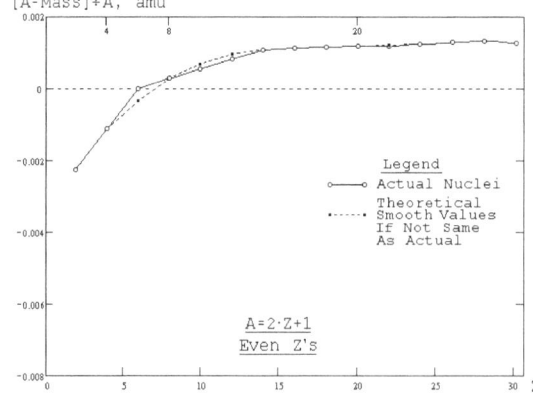

(c) Series s = +1
Figure A-2-12 Even

That those three cases have relatively the smallest overall sizes, are the most compact, is apparent from the relative volumes, relative surface areas and relative inscribed spheres indicated in Figure A-2-11. Figure A-2-13, below, depicts these five polyhedrons to the same scale, that is the same edge length, *"a"* in Table A-2-11. The relative compactness of the three equilateral-triangle-faced polyhedrons is apparent.

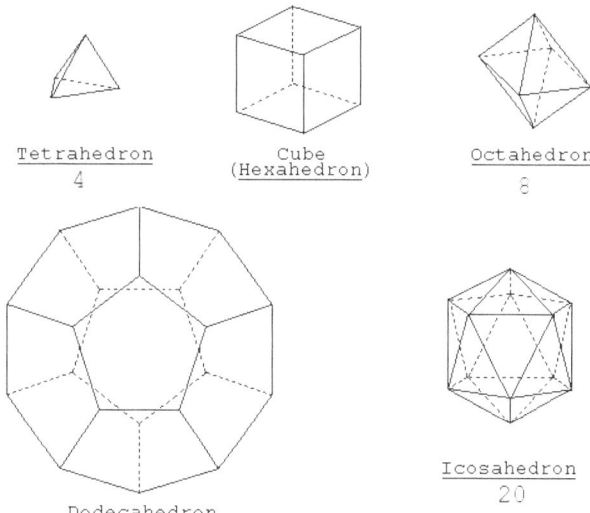

Figure A-2-13
The Regular Polyhedrons [The Platonic Solids]

The relationship between these solid geometric forms and atomic nuclear structure, which relationship would appear to be indicated by the correlation of the "number of faces" of the three most compact of the five regular polyhedrons with the regular dips from the otherwise smooth variation [dashed line] in the mass curves of the low Z, *low* S, atomic species, is as follows.

(1) The theoretical assembly of an atomic nucleus from its component particles involves the assembling together of a number of like charges: a number, N, of electrons and a larger number, A, of protons.

The nucleus being the resultant of A protons and *[N = A-Z]* electrons, the amount of negative charge to be assembled is less than the positive. Consequently, it can be deemed that the assembly is first of the N electrons into a core around which the A protons are then assembled.

(2) In such an assembling of like charges, for example the electrons, the like charges all mutually repel each other with the Coulomb force. Consequently, they automatically space at equal separation distances in the form of a sphere in space. Assembling them into a nucleus is a case of reducing the size of that sphere to the point where the individual particles', *Spherical-Centers-of-Oscillation*, merge.

(3) That configuration in space before the merging is geometrically equivalent to the sphere inscribed inside a regular polyhedron. The center of each face of the polyhedron corresponds to the location of the charges. The inscribed sphere touches each face at just that point.

When the number of merging particles does not correspond to the number of faces in one of the five regular polyhedrons the configuration of the mutually repelling particles is still according to a polyhedron having its number of faces equal to the number of like charge particles that are merging. However, the polyhedron is not regular and that means that the particles are unable to space equally. The best that they can do is arrive at some more or less stable balanced mixture of separation distances that vary around the average value.

The resulting corresponding polyhedron is a quasi-regular form having polygons of various numbers of sides as its faces. It is not as compact as would be the case if it were regular. Its inscribed sphere does not touch all of its faces, only the nearest ones, and that means that some of the charges are radially farther from the center than others.

If the polyhedron corresponding to the assembling charges is regular then the radial distance of each of the charges from the center is the same. The figure is more compact. And, if the polyhedron is of the type having equilateral triangles for its faces, that is a polyhedron of *4, 8,* or *20* faces, representing an assembly of *4, 8,* or *20* like charges, then the radial distance of each charge from the center is a minimum, the configuration is maximally compact.

The more compactly these like charges can fit together the greater will be the potential energy between them and, consequently, the greater will be the energy which they lose for their merging into a new nuclear supercenter to take place. Compactness of the natural configuration of the like charge particles assembling into a nuclear supercenter corresponds directly to the mass decrease exhibited by that nuclear type.

In the graphs of Figure A-2-12, the vertical axis is $[A - Mass]/_A$. Therefore, smaller mass (greater mass decrease) produces higher points on the curves, larger mass (smaller mass decrease) produces dips in the curve. The high points on the curves correspond to greater compactness of the assembly configuration. The dips correspond to less compact cases.

In the assembling of *N* electrons and *A* protons, the *N* electrons and a corresponding *N* out of the total of *A* protons offset each other. Their merger of mutual attraction occurs naturally and readily. Only the excess *Z* protons remaining have the above described configuration problems as they are being assembled into a nuclear supercenter.

That is the significance of the points at *Z = 4, 8,* and *20* in the curves of Figure A-2-12. Some several pages previously, after Figures A-2-10 it was stated that:

> "These data indicate that there is a simple and regular mode of behavior, structure or process that operates effectively for middle and high *Z* or high *s series*, that the variations from nuclear type to type are smooth and regular there. That mode appears to also operate for low *Z*, low *s series*, but is apparently there partially overwhelmed by some other effect not so far detected and taken into account."

That behavior is the assembly configuration effect analyzed and developed above and now "detected and taken into account". Without that phenomenon the variation in mass from nuclear type to type would be completely smooth and regular.

It must be emphasized that there is no contention that the nuclear species actually materially form via the simultaneous combining of *N* electrons and *A* protons. There is no mechanism available to produce such an effect except within intensely hot stars, and even there the combinations effected must be of two particles at a time. The coincidence of simultaneity for combining a greater number of particles at a time is prohibitive.

The effect of assembly configuration that has been presented stems from that the net resulting atomic nucleus, those nuclei as they must materially exist, must have masses as if they had been so constituted. That yields the minimum mass / energy case.

There are no regular polyhedrons of an odd number of faces. The consequence of this geometric condition is that the odd z nuclear species are slightly less compact, have slightly less reduced mass, have slightly greater relative overall masses, and are somewhat less stable or exhibit fewer stable isotopes than their even z counterparts.

HOW THE STABLE NUCLEI CAME TO BE ASSEMBLED

While some of the presently existing atomic nuclei were manufactured in stars, the vast majority of all of the atomic nuclei in the universe are products of the Big Bang. Its initial instant was the equivalent of a pair [particle and antiparticle] of single, very unstable, immense atomic nuclei.

Its "Big Bang" was an explosive nuclear decay from its heavy complex composition through many various stages of multiple less heavy less complex products until ultimately some arrived at stable forms while others, still unstable, decayed further. Some are still present today as long half-life slowly decaying forms.

Whether a particular case was stable, that is optimally compact to a mass minimum per the S-shape or fell into the unstable category, was a matter of mere chance. The S-shape selected the stable ones.

Those decay chains that ended in stable species appear to us as the various stable atomic species of the Periodic Table of the Elements."

THE CAUSE OF THE S-SHAPE

The characteristic *S-shape*, the shape that makes for the stable isotopes amid a sea of unstable ones, comes about as follows.

On the one hand, as the number of electrons in the composition of a nuclear supercenter becomes greater the number of neutrons becomes greater and, consequently the number of multiples of the mass increase of *840 µ-amu* per neutron applied to the nuclear type.

On the other hand, as the number of electrons in the composition of a nuclear supercenter becomes greater the central negative charge attracting the positive protons as a group becomes larger and tends to produce a more compact overall result.

The first tendency is to increase the nuclear mass and the second is to decrease the nuclear mass, both as the N/A ratio increases.

If the ratio is very small, that is if there are few or no electrons in the nuclear composition, then the compactness is quite poor, what with the attempting to combine the mutually repelling protons unaided by a central negative charge.

If the ratio is quite large, that is if the nuclear composition is almost all net neutrons, then the neutron mass excesses overwhelm any small mass decrease due to the few un-neutralized protons, even though they are well compacted.

Only in the range of balance of these two tendencies can a mass minimum be achieved. That occurs at and a little above $N/A = 0.5$ as indicated in Figure A-2-14, below. The Figure is schematic, not precisely quantitative, and only intended to indicate the general form and tendency of the effects.

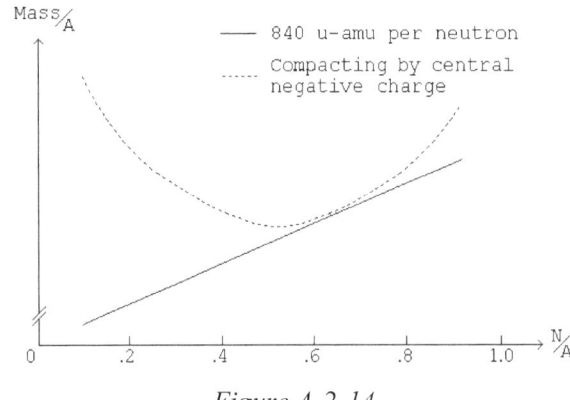

Figure A-2-14

CONCLUSION

The atomic nuclei are each a complex supercenter, a single unitary particle, a *Spherical-Center-of-Oscillation* oscillating as the sum of the oscillations of its components, N electrons and A protons, the oscillation being as presented in equation A-2-2. However, the frequencies in that oscillation are not merely N multiples of the electron rest frequency and A multiples of the proton rest frequency. Rather they are determined by a complex action derived from the theoretical assembly of the nucleus as if from the component particles approaching each other.

The frequency content of equation A-2-2 must correspond directly and exactly with the mass of the nucleus just as the two frequencies in the neutron oscillation wave form correspond directly with the neutron mass. For a nuclear type that matches a regular polyhedron so that the assembling charges can be perfectly equidistant, the frequency of the equation A-2-2 component corresponding to those particles would be A or N multiples of the frequency corresponding to the energy of one of those equidistant and, therefore, equal energy particles.

For the more common case in which the assembling particles are unable to be perfectly equidistant because of the prohibitions of the spherical geometry, the non-matching to a regular polyhedron, the frequency content of equation A-2-2 would be per equation A-2-9, below.

(A-2-9)

$$A \cdot f_p = \sum_{i=1}^{A} [i^{th} f_p] = \sum_{i=1}^{A} [i^{th} m_p] \cdot \frac{c^2}{h}$$

$$N \cdot f_e = \sum_{i=1}^{N=A-Z} [i^{th} f_e] = \sum_{i=1}^{N=A-Z} [i^{th} m_e] \cdot \frac{c^2}{h}$$

The result is that the precise mass of any particular nuclear type depends on the ratio of the number of negatively charged components to the number of positively charged ones and how compactly those charges can arrange themselves overall. The mass of the resulting nucleus is the minimum energy / mass configuration of the charges. The dependency on configurational compactness is attested not only by the natural physical logic of the action but also by the congruence of the especial cases at $Z = 4$, 8, and 20 with the geometry of the regular polyhedrons.

Those of the emerging decay products of the Big Bang that by chance arrived at a maximally compact N/A ratio became our stable elements.

$\longrightarrow$

Appendix A-3

Radioactivity

Radioactivity is an atomic nucleus spontaneously dividing into two (or more) lighter nuclei. Usually it consists of the emission of a relatively light particle and a resulting slightly reduced remaining nucleus. The relatively light particle emitted is usually an electron (- beta), positron (+ beta), helium nucleus (alpha), Hydrogen nucleus (proton), or a neutron. The process appears to be randomly spontaneous for the unstable nuclear specie and does not occur at all for the stable ones.

The reasons for nuclear stability / instability have already been discussed in the preceding section. Nuclei having sufficient mass / energy (*i.e.* having positive separation energy) to make up the decay product particles plus provide their escape energy are unstable. Those with negative separation energy experience stability enforced by the principle of conservation. However, the unstable nuclei do not promptly decay. If they did there would be none of them present now and they would most likely be completely unknown to us. The decay is a process extended in time in various amounts depending on the particular situation.

Characteristic of radioactive decay is that the rate of decay for a particular decay process of any particular nuclear specie continuously declines exponentially according to equation A-3-1.

(A-3-1) $\quad N = N_0 \cdot \varepsilon^{-t/\tau}$

$\quad$ where: N = number of decays during time t.
$\quad$ N_0 = number of nuclei in the sample at time t = 0.
$\quad$ ε = the natural exponential base = 2.718282...
$\quad$ τ = (Greek letter "tau", a constant characteristic
$\quad\quad$ of the particular specie and its decay.

The form of this decay is depicted in Figure A-3-1 on the following page.

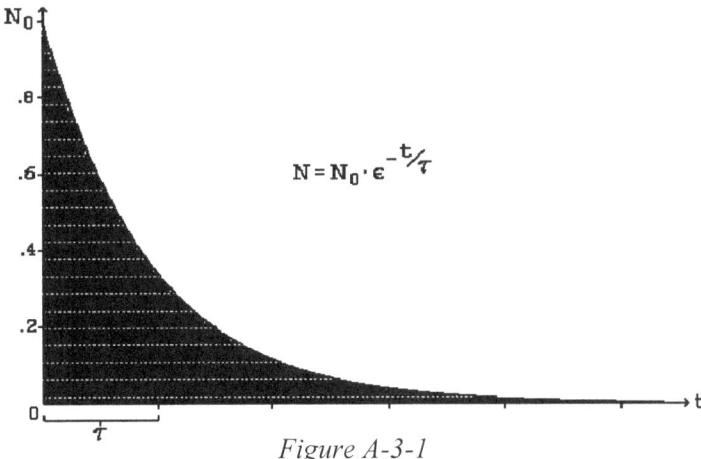

Figure A-3-1

This behavior comes about as follows.

ANALYSIS OF RADIOACTIVITY

As has already been presented, the wave forms of the various nuclear specie, *i.e.* the variation with time of the oscillation of their *Spherical-Centers-of-Oscillation* that are their atomic nuclei, are quite complex. In order for that nucleus to exist as it is at one instant of time and to exist an instant later as a different specie plus a second particle/ specie the transition must be smooth and continuous. And conservation must be maintained.

Those specifications place the following requirements on the oscillation wave forms throughout the transition.

(A-3-2) (1) The wave forms must be smooth:

$$U\left[_{Z}Sym^{A}\right]_{Before} = U\left[_{Z}Sym^{A}\right]_{After} + U\left[\begin{matrix} Emitted \\ Particle \end{matrix}\right]$$

(2) The rate of change of the wave forms must be smooth:

$$\frac{d}{dt}\left[U\left[_{Z}Sym^{A}\right]_{Before}\right] = \frac{d}{dt}\left[U\left[_{Z}Sym^{A}\right]_{After}\right] + \frac{d}{dt}\left[U\left[\begin{matrix} Emitted \\ Particle \end{matrix}\right]\right]$$

(3) The rate of change of the rate of change of the wave forms must be smooth:

$$\frac{d^{2}}{dt^{2}}\left[U\left[_{Z}Sym^{A}\right]_{Before}\right] = \frac{d^{2}}{dt^{2}}\left[U\left[_{Z}Sym^{A}\right]_{After}\right] + \frac{d^{2}}{dt^{2}}\left[U\left[\begin{matrix} Emitted \\ Particle \end{matrix}\right]\right]$$

(4) And so on for all successively higher rates-of change/ derivatives.

All of those specifications were applicable to the origin of the Big Bang treated in Section 1 and, as in that case, they require that the second, newly introduced particle, the "emitted particle" be of the *[1-Cos]* form as

(A-3-3) $\quad U\left[Particle\right] = U_{c}\left[1 - Cos\left[2\pi \cdot f_{Particle} \cdot t\right]\right]$

126

The oscillation of the decaying nucleus is of the form

$(A-3-4)$ $\quad U\left[_{Z}Sym^{A} \right] = A \text{ protons} + [N = A - Z] \text{ electrons}$

$$= U_{c} \cdot \left[A - \cos(2\pi \cdot [A \cdot f_{p}] \cdot t) \right] + \left[-U_{c} \cdot \left[N - \cos(2\pi \cdot [N \cdot f_{e}] \cdot t) \right] \right]$$

$$= U_{c} \cdot \left[Z - \cos(2\pi \cdot A \cdot f_{p} \cdot t) + \cos(2\pi \cdot N \cdot f_{e} \cdot t) \right]$$

and the above specifications also require simultaneously:

> that the first instant of the new particle appear at a point in the overall oscillation of the radioactively decaying nucleus at which the *N multiple electrons* portion is at zero and just beginning a cycle of its oscillation;

> and that the first instant of the new particle appear at a point in the overall oscillation of the radioactively decaying nucleus where the *A multiple protons* portion is at zero and just beginning a cycle of its oscillation.

Those requirements present a real problem because the specified *zero points* in each of the two different freqency's oscillations are rarely simultaneous and most likely never occur simultaneously. That is because both frequencies are always irrational numeric quantities and unable to relate as one being an integer multiple of the other. That is because both numeric values depend on the Planck Constant, h, which itself is irrational, equation $A-3-5$,

$(A-3-5)$ $\quad f = m \cdot c^{2} \Big/ h$

where m is the kinetic mass of the nuclear component multiple proton or electron.

While the oscillations' pairs of *zero points* do not achieve exact simultaneity they do approach it. The relative phase of the two oscillations, one at $A \cdot f_{p}$ and one at $N \cdot f_{e}$, is continually shifting and from time to time may become such that at some particular point the *zero points* are very close. Such closeness might occur every few seconds, or minutes or essentially regularly at longer intervals. At the lowest frequency / longest wavelength that could be involved, that of a single electron $f_{e} = 1.235,589,965 \cdot 10^{20}$ hz, there are that many electron *zero points* per second.

But, the oscillations of the various atoms of the element that is undergoing radioactive decay are not synchronized. They do not approach a pair of *zero points* at the same time. Rather their relative phases are distributed essentially uniformly over the range of relative points along the complex pattern of oscillation, equation $A-3-4$, at which they could be at any particular instant of time.

That is because unless they are at absolute zero temperature and energy all particles are in motion. Their oscillation pattern depends on their velocity, its speed and its direction. These particles regularly emit and absorb photons of radiation – the usual Rayleigh-Jeans or black body radiation activity. The particles' energies are continuously changed in consequence. That means that their velocities change and thus their oscillation patterns change.

A nucleus' *Separation Energy* includes energy in excess of that to account for the separated masses. So long as overall energy and momentum are conserved the requirements of equation $A-3-2$ are somewhat flexible in the allocation of increments of the energy and momentum of the nucleus before decay partly to the remaining nuclear specie after the decay, partly to the emitted particle, and partly to an optional gamma, neutrino or whatever that may be radiated. Those energy increments correspond to minor variations in the particles' oscillation patterns, equation *A-3-5*. Thus the satisfaction of equation *A-3-2* can occur over a limited range of points that are in themselves very near to being a close pair of *zero points* in the oscillation pattern. An exact "match" to one single perfect *zero point* is not required because the variations in acceptable energies create some leeway.

There is a range of states, a "decay window", a small group of points along the total nuclear oscillation over its total period, within which the requirements of conservation and of equation *A-3-2* can all be satisfied. If the actual oscillation passes into any of the states within the "window" then the decay occurs.

At any instant of time a few individual nuclei are just a moment away from entering a state within a "decay window", some are within the "window", some are just leaving such a state. The vast majority are uniformly spread over the range of states between the "decay windows". If the average time between the beginnings of decay windows, the average duration before entry into a decay window, is referred to as τ then during a minute time interval Δt the fraction of the total number of nuclei in the sample that will progress in their oscillation pattern to the point of being in a decay window state is $\Delta t/\tau$. If there were no other factors affecting the process then the same fraction, $\Delta t/\tau$, would decay each Δt.

However, within that sample each nuclear *Spherical-Center-of-Oscillation* is continuously interacting with incoming *Propagated Outward Flow* of other centers and photons. The status of its oscillation is continuously being changed by those encounters and the consequent changes in energy and motion of the nucleus. For the purpose of determining where a particular nucleus is along its pattern of oscillation from one "decay window" to the next, each of the nuclei are continuously being shuffled and reshuffled, distributed and redistributed over the range of possibilities, effectively randomly, and effectively uniformly.

Thus the state of the above sample progresses as follows.

```
(1) Uniform distribution of states.

(2) In the next Δt fraction Δt/τ progresses in their
    oscillation pattern to the point of being in a decay
    window state and decays (It is a minute fraction since
    Δt is minute).

(3) The states of the undecayed fraction, [1 - Δt/τ] of the
    original total, are redistributed uniformly.

(4) Fraction Δt/τ of that undecayed remainder decays.

(5) The states of the undecayed fraction, [1 - Δt/τ] of the
    then undecayed remainder, again are redistributed
    uniformly,

(6) And so on.
```

This process yields an exponential decay as presented in equation *A-3-1* as follows.

(A-3-6)   N ≡ the number of undecayed nuclei in the sample.
dN ≡ the change in the number of undecayed nuclei
during infinitesimal time interval, dt.

Number of Decays in dt as Fraction of $N = \dfrac{\Delta t}{\tau} \to \dfrac{dt}{\tau}$ as $\Delta t \to 0$

$$dN = -N \cdot \frac{dt}{\tau}$$

$$\frac{dN}{N} = -\frac{1}{\tau} \cdot dt$$

$$Ln\,[N] = -\frac{1}{\tau} \cdot t + C$$

$$N = N_0 \cdot \varepsilon^{-t/\tau} \quad \text{which is equation } A-3-1$$

HOW THE RADIOACTIVE DECAY TAKES PLACE

A typical case is the radioactive decay of *Helium 6* into *Lithium 6* by the emission of a *-beta electron* and a *neutrino* [η] carrying off excess energy.

(A-3-7) $U\left[\,_2He^6\right] \;\Rightarrow\; U\left[\,_3Li^6\right] \;+\; U\left[Electron\right] \;+\; \eta$

$$= U_c \cdot \left[2 - Cos(2\pi \cdot 6 \cdot f_p \cdot t) + Cos(2\pi \cdot 4 \cdot f_e \cdot t)\right]$$

$$\Rightarrow U_c \cdot \left[3 - Cos(2\pi \cdot 6 \cdot f_p \cdot t) + Cos(2\pi \cdot 3 \cdot f_e \cdot t)\right]$$

$$- U_c \cdot \left[1 - Cos(2\pi \cdot f_e \cdot t)\right] \;+\; \eta$$

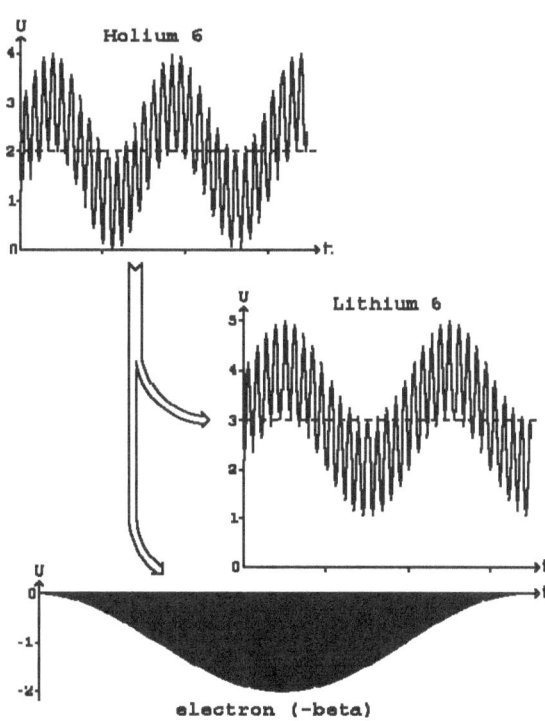

Figure A-3-2
Radioactive Decay of Helium 6 into Lithium 6 and an Electron

129

Initially from the above equation A-3-7 and Figure A-3-2 it would appear difficult if not impossible to visualize or understand how the actual real world separation of the pre-decay nucleus into the decay products can take place. However, examining equation A-3-7 more closely it can be seen that the only change is in the multiple electron portion of the overall nuclear oscillation:

Before the decay the multiple electron portion of the overall oscillation is as follows. The change is the removal of one electron from the four.

(A-3-8) $-\left[4-\mathrm{Cos}(2\pi\cdot[4f_e]\cdot t)\right] \Rightarrow$

$$-\left[-\left[1-\mathrm{Cos}(2\pi\cdot[1f_e]\cdot t)\right]\right] =$$

$$-\left[3-\mathrm{Cos}(2\pi\cdot[3f_e]\cdot t)\right]$$

The cosine function is the same as the following infinite series.

(A-3-9) $$\mathrm{Cos}(x) = 1 - \frac{x^2}{2!} + \frac{x^4}{4!} - \frac{x^6}{6!} \ldots$$

Expressing the multiple electron portion of equation A-3-8 in terms of the cosine infinite series the result is equation A-3-10. [The factors 2π and t of each cosine argument are omitted for clarity, e.g. $2\pi\cdot[4\cdot f_e]\cdot t$ is rendered as $4f_e$.]

(A-3-10)

$$\left[4 - \mathrm{Cos}(4f_e) = 4 - 1 + \frac{4f_e^{\;2}}{2!} - \frac{4f_e^{\;4}}{4!} + \frac{4f_e^{\;6}}{6!} \ldots\right] \Rightarrow$$

[To avoid clutter e.g. $[4\cdot f_e]^2$ is done as $4f_e^{\;2}$]

$$-\left[1 - \mathrm{Cos}(f_e) = 1 - 1 + \frac{f_e^{\;2}}{2!} - \frac{f_e^{\;4}}{4!} + \frac{f_e^{\;6}}{6!} \ldots\right] =$$

$$\left[3 - \mathrm{Cos}(3f_e) = 3 - 1 + \frac{3f_e^{\;2}}{2!} - \frac{3f_e^{\;4}}{4!} + \frac{3f_e^{\;6}}{6!} \ldots\right]$$

There the process of particle separation and emission and the process of change in the remaining nucleus become clear.

Because the satisfaction of equation A-3-2 can occur over a limited range of points that are in themselves very near to being a close pair of *zero points* in the oscillation pattern the failure of the decay to occur at an exact "match" to one single perfect pair of *zero points* accounts for the necessity of the neutrino product particle, the η of equation A-3-7.

THE NEUTRINO

- Neutrinos have been detected at energies of from *1 to 10^{17} eV*, on the order of *10^{-9} to 10^{8} amu equivalent.*

- They have no electric charge.

- They have an extremely small interaction with matter. That is, a neutrino can pass through an immense amount of matter with no apparent interaction (e.g.

the vast majority of solar neutrinos that encounter planet Earth simply pass completely through the Earth with no interaction.)

- The neutrino appears to be a "particle" with an extremely small rest mass equal to about one ten-thousandth, 0.0001, of an electron rest mass [otherwise by far the smallest].

- The neutrino is produced by radioactive decays.

- The options as to the form of the neutrino are: (1) A new type of *Spherical-Center-of-Oscillation* or (2) a kind of electromagnetic modulation of a center's *Propagated Outward Flow* caused by the center's motion, i.e. a form of photon.

Photons are generated by orbit changes of atomic orbital electrons. The orbital electrons have angular momentum and that angular momentum changes when the orbit changes so that the photon emitted carries off the change in the angular momentum of the electron.

Neutrinos, on the other hand, are generated by radioactive decay of non-orbiting atomic nuclei that do not have the angular momentum that atomic orbital electrons have. Whatever momentum neutrinos carry away from a nuclear decay is linear momentum, not angular.

Photons ranging from radiant heat to gamma, γ. rays occur over the same entire range of energies as do neutrinos. There is no accommodation for neutrinos in the family of photons.

The neutrino, then, is a new type of *Spherical-Center-of-Oscillation* of extremely small rest mass / rest energy and zero average oscillation value [no electric charge], $\eta = Cos(2\pi \cdot f_\eta \cdot t)$. The f_η is the equation $A-3-5$ equivalent of the kinetic mass m_η of the neutrino.

The detection of neutrinos is a matter of detecting changes that they produce in encountered orbiting electrons. But, because the neutrino lacks the angular momentum necessary to be delivered to an orbiting electron to cause it to change orbits the neutrino seldom produces such a change. Now and then, but rarely, the circumstances may allow a neutrino-caused electron orbit change. The circumstances would be a neutrino encountering an electron where the spatial relationship of the neutrino's linear momentum to the electron's orbit were such that the neutrino need supply only a linear momentum change to achieve the effect of an angular momentum change on the electron.

"STRANGE" PARTICLES

With the development of increasingly greater energies available in particle accelerators physics researchers have discovered an increasing number of particles.

Acknowledging that the analogy is somewhat brutal, nevertheless that research in high energy physics is not unlike the study of the composition and fundamental parts of a limousine by hurling everything from roller skates to motorcycles at it with as much energy as possible and then analyzing the resulting pieces.

It is true that little alternative seems to be available for experimental procedures to study the atomic nucleus, but when the magnitude of the disruptive energy used to generate the pieces is considered taking the resulting "pieces" seriously as a key to the nature of matter makes limited sense. Furthermore, those "pieces" may not be so much fundamental "building blocks" of matter as the fragments into which *Spherical-Centers-of-Oscillation* naturally break under such energies, so to speak a reflection of the "fault lines" in the center's oscillations.

Of all of the many particles discovered in high energy physics research:

- Those having rest mass are *Spherical-Centers-of-Oscillation* (pieces that are centers smashed out of existing centers);

- Those having charge have oscillations with a non-zero average level;

- Those having momentum have a non-spherically symmetrical oscillation, the "axis" about which the oscillation remains symmetrical pointing in the direction of the momentum;

- Those not having rest mass are brief fluctuations in the *Propagated Outward Flow* from *Spherical-Centers-of-Oscillation*.

$\longrightarrow$

$\longrightarrow$

134

Appendix A-4

The Photon

THE PROBLEM

The problem of the photon is how to resolve that light exhibits behavior only explainable if it is an electro-magnetic [*E-M*] wave phenomenon yet light also exhibits behavior that would appear to be only explainable if it is a particle.

The evidence for the wave nature of light is extensive including the wave behaviors of: reflection, interference, refraction, diffraction, frequency, wavelength, and polarization as well as the highly successful Maxwell's Equations.

Then phenomena appeared that seem to require a particle nature of light hence its particle name, "photon". Those phenomena were the failure at short wavelengths of the theoretical Rayleigh-Jeans law of black body radiation, the photoelectric effect, and the line spectra of gases.

This evidential wave-particle duality led to the concept of the photon as a particle in the form of a "wave packet". But, the particle nature of the photon still has a number of problems.

A wave in free space spreads out as it propagates, but the particle photon "wave packets" must be considered as staying together like a particle. The *E-M* wave front is continuous, but a front of propagating particles involves the particles' moving radially from the source with the distance between particles increasing with distance from the source and nothing in the spaces between.

E-M radiation is produced by acceleration of charge and must produce *E-M* propagation that is spatially symmetrical to the charge's motion, but the particle theory requires that the radiation travel away from the accelerated charge as a specific particle in some specific direction without symmetry.

The next particle may be in another direction, the next in a third, and so on, so that a large number of radiated particles exhibit a dispersion pattern like that of the wave field, but that still is behavior that is inconsistent with the wave aspect.

ANALYSIS OF THE PHOTON FROM ITS GENERATING SOURCE

To resolve this problem it is necessary to go first to the constraints on what a photon is as they are imposed by its source, the transition of an atomic orbital electron from an outer to an inner orbit which transition must fit and match to, the following requirements.

- The transition is a change from the initial state to the final state as in part of a single cycle of an oscillation. It is not a changing to a different state and then returning back to the original state as in a full cycle of an oscillation.

- To avoid an infinite rate of change, which is impossible, the transition must be a smooth variation, without any sudden "jump".

- The resulting radiation exhibits all of the characteristics of Maxwellian electro-magnetic wave and is at one simple frequency, the photon frequency. It therefore must be in the form of a simple sinusoid.

- The theory of information in communications shows that at least a sample every half-cycle of an oscillation is required to specify it sufficiently. Therefore, at least a half cycle of the photon oscillation is required to specify it.

For all of the above reasons, the photon must be in the form of a half-cycle of a sinusoidal function of time.

- Magnetic field is directly proportional to the velocity of the moving electric charge producing that field. Therefore the magnetic field of the photon is directly proportional to the transitioning electron's velocity.

  Since the photon magnetic field must be a half-cycle sinusoid the transitioning electron's velocity variation must be of a half-cycle sinusoid form.

- The electron velocity must vary in accordance with the above from the stable velocity of the initial orbit through a period of increase and ending in the stable velocity of the final orbit.

[The potential energy lost in the move to a lower orbit appears half in the emitted *E-M* radiation, the photon, and half in the increase in electron kinetic energy due to its greater velocity in the inner orbit].

The combination of these factors results in the specification that the photon must be a half cycle of a pure sinusoidal type variation behaving as in the following figure.

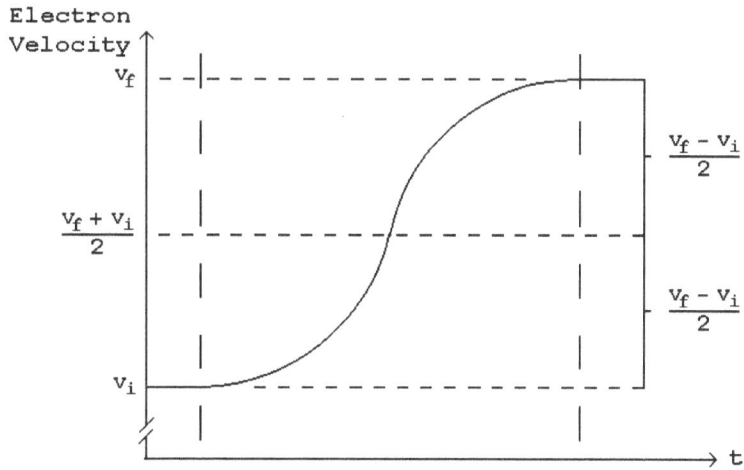

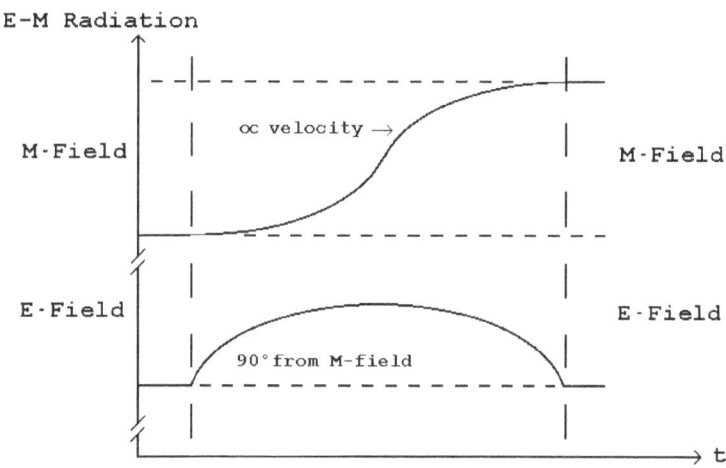

Figure A-4-1
The Orbit Transitioning Photon Generating E-M Behavior

As the transitioning electron, a *Spherical-Center-of-Oscillation*, follows its transition path, Figure 2 on the following page, the above form of half-cycle sinusoid is imposed on its *Propagated Outward Flow* as a half-cycle burst of modulation of that flow functioning as a carrier wave for the modulation.

The electron, traveling from its initial outer orbit to its final inner orbit with its velocity gradually increasing in a sinusoidal manner as in Figure 1, follows a path as illustrated in Figure 2, below, and emits an *E-M* wave field in "doughnut form" as in Figure 3, below, relative to its instant-by-instant varying vector velocity direction at each instant of the transition.

The peculiar shape of that field because of the directional orientation of the "doughnut" swinging through a substantial portion of a full circle according to the path of the electron's orbital descent causes the propagated *E-M* wave to contain the requisite form, angular momentum and energy for causing an encountered orbital electron

137

elsewhere to be elevated to a higher orbit equivalent to the higher orbit that the electron previously descended from. The propagated *E-M* burst contains and transmits both energy and angular momentum.

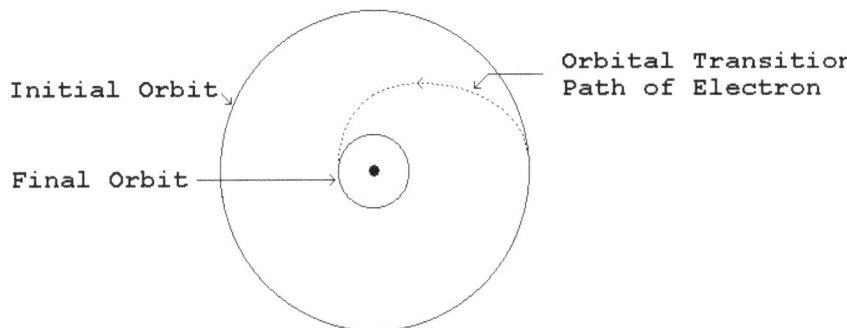

Figure A-4-2
Typical Electron Outer-to-Inner orbit Change Path

In terms of the final orbital period the transition takes place in $1\frac{1}{3}$ orbital periods at most and in approximately ½ or less such orbital periods for most cases. See *Notes re Orbital Transitions*, further below.

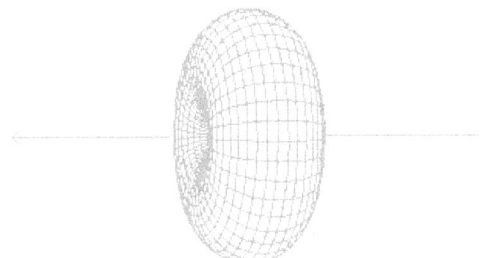

Figure A-4-3
Instantaneous Electro-Magnetic Radiation Pattern [doughnut shape]
of a Single Electron Traveling Horizontally.

It is clear from the above that for the radiation emitted in this circumstance, which radiation is the photon, it is impossible that it travel outward away in a single specific direction as required for the photon-particle hypothesis. Furthermore, there are other problems with that hypothesis.

Perhaps the greatest other problem with the photon-particle theory is as follows. The wavelength of light is in the range of 10^{-7} meters . Atomic dimensions are on the order of 10^{-10} meters so that if a photon is to contain wavelength data relevant to the light that it represents, it must then have dimensions that are on the order of $10^3 = 1000$ times the size of an entire atom let alone than the size of a much smaller orbital electron. Clearly this is completely at variance with the photon-particle explanation of the photoelectric effect and line spectra of atoms.

For example, in the instance of absorption of a photon-particle causing the raising of an orbital electron to a higher orbit there would be, relatively speaking, a *football size photon* interacting with a *sand grain size atom*, the football-photon

138

managing to focus its action solely on one *germ size electron* in the *sand grain size atom* without disturbing any of the rest of the atom.

Conversely, in the instance of an orbital electron falling to a lower orbit and emitting a photon-particle to carry off half the lost energy, there would be a *germ size electron* giving birth to a *football sized photon*.

The photoelectric effect has the same problem. For a photon-particle to eject an electron from an encountered metal material means a *football size photon* interacting with a *germ size electron* in the material composed of *sand grain size atoms* without disturbing anything other than the particular *germ size electron*.

RESOLUTION OF THE PHOTON'S SEEMING PARTICLE-LIKE BEHAVIOR

The description of the earlier above scenarios, there illustrated in terms of relative metaphorical particles [*football photon, sand grain atom, germ size electron*], however now treated as the interaction of *E-M* wave radiation and the atomic orbital electron, is as follows.

The atomic orbital electron falling from an outer orbit to an inner orbit follows a curved path somewhat less than a full orbital pass. Throughout the duration of that curved descent path the electron gradually loses energy and emits *E-M* wave radiation <u>outward in all directions including at right angles to its path</u> [emits its *Propagated Outward Flow* as modulated by the in-process-of-being-generated photon] in a pattern radially symmetrical to the instantaneous vector direction of the electron at any moment. That brief burst of a half cycle of *E-M* radiation carrying and transmitting orbital change energy and angular momentum, propagates on outward decreasing in amplitude as the inverse square of distance from its point of origination.

Such radiation bursts encounter other atoms orbital electrons at various small parts of the radiation bursts' overall wavefront periphery at various distances from their source. Sometimes at such an encounter, out of the arriving plethora of such bursts, a subset arrives coordinated enough to match the energy, frequency and angular momentum requisite to elevate the encountered electron from its current stable orbit exactly to a higher stable orbit through an increase in its energy and angular momentum by the amount in the combined bursts in the subset of the incident *E-M* waves. That is a wave event in spite of that it is a coordinated subset of minute regions of wavefront periphery of the various incident waves giving it the "look" of a particle.

Similar action takes place in the photoelectric effect, the difference being that in the photoelectric case photon burst subsets of sufficient energy elevate electrons to an electric current free of the metal atom on which the incoming photon was incident. That "incoming photon" is actually a coordinated subset of the plethora of half-cycle *E-M* wave bursts. All arriving natural "photons" are such.

Of the total of all individual wave bursts that directly encounter a particular orbital electron most of the time the effect is not to elevate the electron to a higher stable orbit. Rather, it is to elevate the electron to a slightly higher energy level where it is in an unstable position and immediately re-radiates the *E-M* wave that acted on it and returns to its stable orbit. To an external observer it is as if nothing happened.

But, a correctly coordinated subset of incoming wave bursts will occasionally elevate an encountered electron to a higher stable orbit producing an absorption spectrum line and an electron in a position from which, if it is disturbed from its new stable orbit, it will radiate a radiation spectrum line.

In all of the radiative cases, whether from orbital electrons or vibrating molecules' Black Body Radiation or other charged particle sudden accelerations, the energy is <u>exchanged</u> in a quantized amount $W = h \cdot f$ as proven by Planck in his resolution of the Rayleigh-Jeans law of black body radiation. But that quantum of energy travels off away from the location at which the initial energy exchange took place as an *E-M* wave propagating outward in many directions. The concentration of the overall total exchange energy rapidly diffuses in space as the surface of the outward propagating wave front increases,

In all of the absorptive cases, orbital electrons or photoelectrons, or Black Body radiated heat *E-M* waves, the arriving and acting "photons" are timing, phase and vector coordinated subsets out of the plethora of half-cycle *E-M* wave bursts, the appropriateness of the subset's composition depending on the particular environment on which it is incident: a particular electron in a particular atomic orbit, a loosely bound electron in a metal surface, an encountered molecule.

But, why is the "photon" energy, the energy in the <u>exchange</u>, dependent only on the frequency as in $W = h \cdot f$? how does wave amplitude enter into the process ?

As they propagate outward the waves disperse so that the amplitude at points on the periphery of each individual burst decreases steadily in inverse square manner from its value at the moment of the interaction that created it. The farther that an absorptive interaction is from the source of incoming radiation the greater the number of individual bursts, each contributing a small portion of the "requisite amplitude", that are required for an absorptive interaction to be able to take place.

If a "brand new" half-cycle burst at full amplitude as just created and before propagation outward were to encounter an electron in the identical energy-momentum-orbit situation as that of the electron that just finished generating the new burst, the new burst would elevate that "just finished" electron back to where it started from. That new burst's before propagation amplitude is the "requisite amplitude" that the actual numerous bursts of the subset must add up to.

The amount of energy naturally depends on frequency. The higher the frequency the more rapidly the *E-M* radiation oscillates. The *E-M* radiation carries the ability to cause corresponding change in motion in encountered charged particles. It requires more energy per time to make a rapid change than to make a gradual one. A shorter period (higher frequency) half-cycle sinusoid must contain directly proportional greater energy to produce the proportionally more rapid change.

SPECIAL CASES - LABORATORY PHOTONS

The discussion and treatment of photons to this point has been of only natural world photons behaving in their natural fashion without any interference or action by humans. As to be expected the situation is somewhat modified when dealing with the effects of humans; however, there is no change in the fundamentals that photons are a purely *E-M* field effect not that of a discrete particle.

Radiating photons as above, whose *E-M* field disperses widely in nature, are focused, collimated into a monodirectional beam in the laboratory. That beam, in its effect upon encountering matter, is effectively a stream of photon particles, each still a half cycle burst of *E-M* field but having a particle's one specific direction.

The Compton Effect

The Compton Effect is considered the *sine qua non* of the particle theory of the photon. When x-rays of known wavelength impact atoms the x-rays are scattered through an angle Φ and emerge at a greater wavelength, meaning reduced frequency, related to Φ. That reduced frequency means reduced energy $W = h \cdot f$.

In the study of the Compton Effect the incoming x-ray radiation is considered to consist of particles and the study analyses a collision of a photon with an electron just as if two billiard balls were to collide in a glancing manner (Figure 4, below). The classical particle physics of the situation require that energy be conserved and that momentum be conserved independently in all directions.

The incoming incident x-rays are, of course, a collimated monodirectional beam of natural photons.

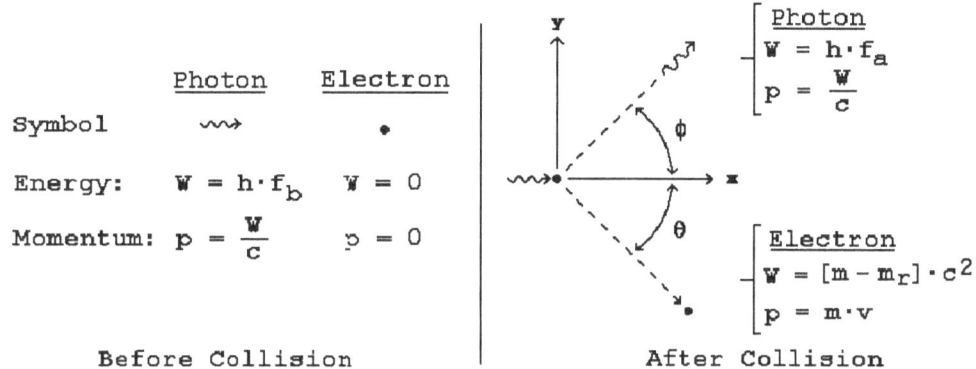

Figure A-4-4
The Compton Effect

Experiment has verified that in fact the scattered radiation is at reduced frequency and in agreement with the Compton developed formula. That has been taken as confirmation of the particle nature of the photon but it is not. It is only confirmation that, of course, energy and momentum must be conserved in such interactions. A particle form of radiation is not necessary to accomplish that.

In the "scattering" in the Compton Effect incoming *E-M* waves are absorbed by the electrons which then radiate new *E-M* radiation. The incoming photons are not particles but the half-cycle *E-M* wave bursts already developed.

Lasers

Lasers are another example of the use of single frequency focused beams of the half cycle sinusoidal bursts of *E-M* waves that would otherwise widely disperse. In the case of the laser the result is an optical "current" of those bursts, which in its single frequency and *E-M* oscillation is essentially equivalent to the electrically generated carrier waves which carry radio and broadcast transmitted information in modulation applied to the carrier.

141

Other Experiments

In general all experiments on the nature and behavior of light depend on using artificial light, that is single frequency light that is focused or collimated into a narrow monodirectional beam, which gives it some of the appearances of a particle while still retaining its *E-M* nature. Thus, in a sense, the problem of the wave-particle duality of light is the issue of "which light" – light as in free nature or specialized laboratory light.

NOTES RE ORBITAL TRANSITIONS

Knowing the time duration of the electron's orbital transition relative to the electron orbital period is helpful in visualizing the process. The orbital transition takes place in the time of one-half cycle of the photon's frequency. Letting "D xx" symbolize "the duration of xx" then

(A-4-1)
$$\text{D orbital transition} = \text{D photon} = \frac{1}{2 \cdot \text{photon frequency}}$$

Let n be the orbit number, an integer representing the number of matter wavelengths in the orbit. The electron orbital velocity is proportional to $1/n$. Its matter wave frequency is proportional to $1/n^2$. The photon frequency is equal to one-half the difference between the initial and final orbits matter wave frequencies. From those, the duration of the orbital transition in terms of the duration per orbit in the final orbit is

$$\frac{\text{D orbital transition}}{\text{D final orbit}} = \frac{n_{initial}^2}{n_{final}[n_{initial}^2 - n_{final}^2]}$$

n_i	n_f	Transition/ final orbit	n_i	n_f	Transition/ final orbit
2	1	4/3	4	3	16/21
3	1	9/8	5	3	25/48
3	2	9/10	6	3	36/81
4	2	4/6	7	3	49/120
5	2	25/42			

Figure A-4-5
Orbital Transitions Parameters

The transition takes place in $1\text{-}1/3$ final orbit periods at most [*where n_i = 2, n_f = 1*], and in less than *1* final orbit period for most cases [*n_f > 1*]. The orbit transition and photon emission take place with the electron traveling a significant portion of an orbit around the atomic nucleus in all cases.

Looked at another way, the transitioning electron travels in the range of on the order of *180º* to *360º* of a circular orbit. Its path being curved, its direction at instants during the transition changes by on the order of *180º* to *360º*. Such a path is completely incompatible with a particulate photon being emitted in some one specific direction by that transition.

$\longrightarrow$

$\longrightarrow$

Appendix B

The Limitation of the Original Envelopes

This is to show how the otherwise infinite string of envelopes to the original oscillation at the start of the universe was subject to a finite limitation. By "finite limitation" is meant that in the vicinity of the cut-off number of envelopes, N_0, the amplitude of each of the further successive envelopes being imposed on the original $U(t)$, equation 2-5 was successively significantly less than its immediate predecessor and the rate of that amplitude decrease increased sharply with further envelopes – there was a sharp cut-off of amplitude.

After a moderate number of such cut-off region envelopes the amplitude of any further envelopes becomes infinitesimal. While such infinitesimal (and still continuing to become ever more infinitesimal) envelopes theoretically go on to an infinite number of them, the result is equivalent to the convergence to a finite value of a mathematical infinite series such as, for example that of the cosine. The envelopes cut-off is a result of the mathematics of $U(t)$.

The key to that behavior is to be found in Table B-1, below, the expansion of the $Cos^n(x)$ function. The "Cosmic Egg" expression, equation 2-5, repeated below

$$(2\text{-}5) \qquad U(t) = \pm U_0 \cdot \left[1 - Cos\left[2 \cdot \pi \cdot f_{env} \cdot t \right] \right]^{N} \cdot \left[1 - Cos\left[2 \cdot \pi \cdot f_{wve} \cdot t \right] \right]$$

contains the factor

$$(B\text{-}1) \qquad Cos^{N_0}\left[2\pi (f_{env}) t \right]$$

which creates the set of envelopes to the original oscillation. The expansion of the cosine raised to the power of its N_0 exponent behaves according to the pattern illustrated in Table B-1, below. Analysis of the patterns in the coefficients of the individual terms of the $Cos^n(x)$ expansion discloses a pattern related to the binomial expansion as demonstrated in the table.

145

(a) Binomial Expansion Coefficients [a + b]^n

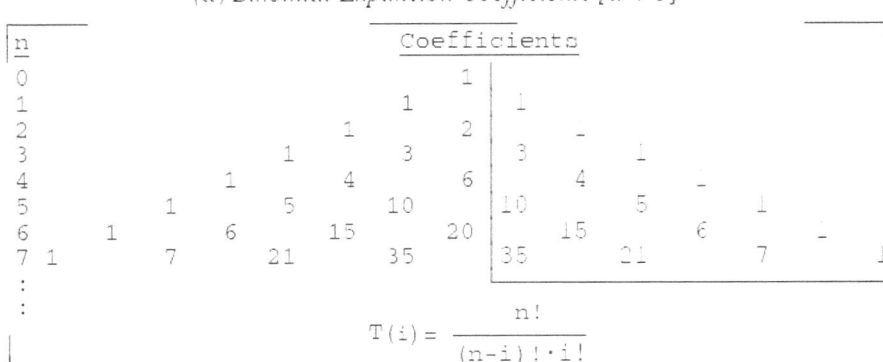

n	Coefficients							
0				1				
1			1	1				
2		1	2	1				
3	1	3	3	1				
4	1	4	6	4	1			
5	1	5	10	10	5	1		
6	1	6	15	20	15	6	1	
7	1	7	21	35	35	21	7	1

$$T(i) = \frac{n!}{(n-i)! \cdot i!}$$

(b) Cos^n(x) Expansion Coefficients

n	Coefficients							
Times Cos(*), * =	0x	1x	2x	3x	4x	5x	6x	7x
0	1							
1	-	1						
2	1	-	1					
3	-	3	-	1				
4	3	-	4	-	1			
5	-	10	-	5	-	1		
6	10	-	15	-	6	-	1	
7	-	35	-	21	-	7	-	

$$T(i) = \frac{n!}{(n-i)! \cdot i!}$$

Table B-1

Clearly, with the exception of the constant term (where, in the table, $* = 0x$) the other terms of the expansion of $Cos^n(x)$ have the same coefficients as the corresponding terms of the binomial expansion. The formula for the binomial expansion can thus be used to obtain the coefficients for any value of n in the expansion of $Cos^n(x)$. in the present case for any value of N_0 in the expansion of the $U(t)$ factor

$$Cos^{N_0}\left[2\pi(f_{env})t\right]$$

The cut-off occurs around the value of N_0 regardless of what that value is. Therefore the value of N_0 is not important. Nevertheless it is of interest that various attempts to estimate it give values around 10^{85}.

$N_0 = 10^{85}$ is the n of the formula. It is not practicable and most likely not possible to calculate all of the coefficients of the cosine expansion of the envelopes for 10^{85} envelopes. On the other hand, it is not unreasonable to calculate the 85 cases corresponding to the frequency multiples of the expansion: 10^1, 10^2, 10^3, $\cdots$ 10^{85}.

Figure B-1, below, is a plot of the relative magnitude of the successive coefficients of the various frequency multiples $(1 \cdot x, \; 3 \cdot x, \; \cdots \; 10^{85} \cdot x)$, in the expansion of $Cos^n(x)$ for $n = N_0 = 10^{85}$. The plot indicates a sharp cut-off, an attenuation of the higher frequencies. Figure B-1(a) uses a linear horizontal axis and shows the cut-off in detail. Figure B-1(b) uses a logarithmic horizontal scale to better present the tremendous range in frequency multiples from 1 to 10^{85}. It shows that the cut-off is quite sharp and drastic.

This cut-off is merely the action of the mathematics of $cos^n(x)$.

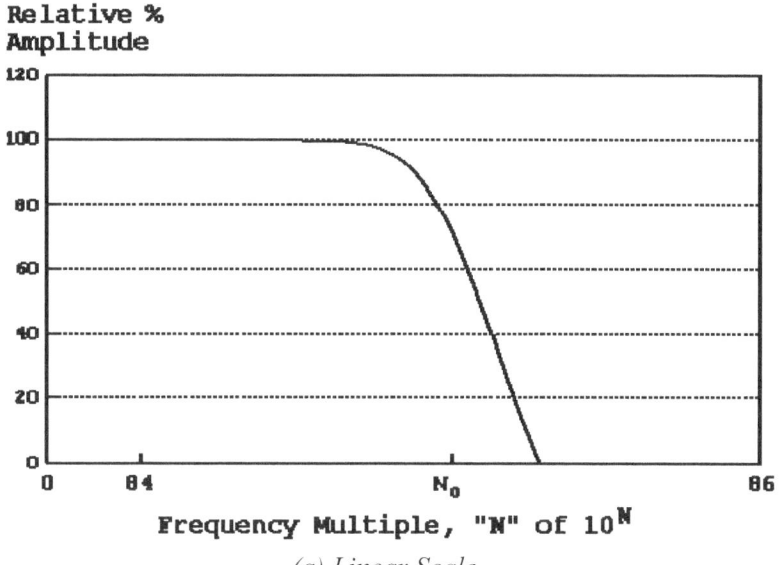

(a) Linear Scale

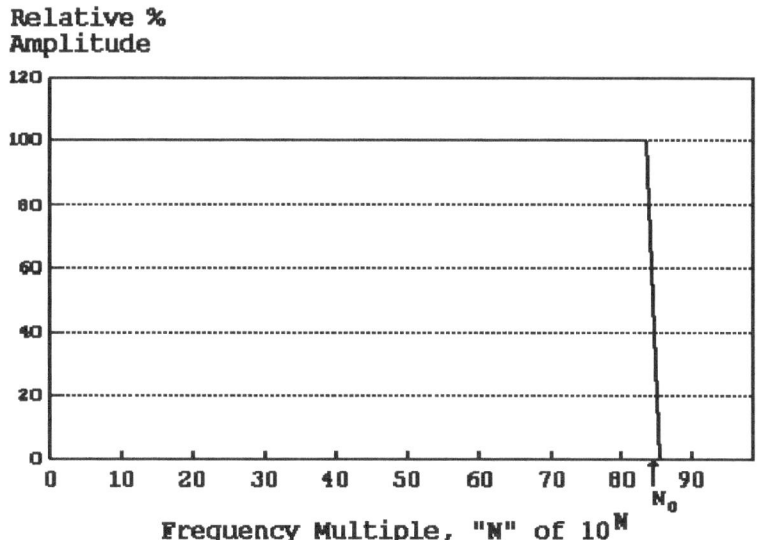

(b) Logarithmic Scale

Figure B-1
The Cos^n(x) Limitation of the "Cosmic Egg

147

$\longrightarrow$

Appendix C

Why No Immediate Mutual Annihilation

BACKGROUND OF THE PROBLEM

The Big Bang could only have resulted in equal amounts of matter and antimatter for the sake of the principle of conservation as presented in Section 1, *The Origin of Matter - Its Cause* with the assumption that there would have been a complete and almost instantaneous mutual annihilation.

Because that annihilation did not take place it has been hypothesized that the original symmetry was slightly skewed in favor of matter and that the universe is now all matter, all original antimatter having been annihilated with an equal amount of original matter. However that skewed balance conflicts with conservation in the Big Bang.

The Big Bang had to produce equal amounts of matter and antimatter and their total mutual annihilation did not occur because of the conditions there. Rather, while a moderate amount of initial matter / antimatter mutual annihilations may have taken place our present universe contains the remaining matter and antimatter in equal amounts, between some particles of which further mutual annihilations still occur at a modest rate.

The failure of comprehensive matter-antimatter immediate annihilation to occur develops as follows.

CONDITIONS AFFECTING MATTER / ANTIMATTER MUTUAL ANNIHILATION

What Is a Matter / Antimatter Annihilation ?

A positron-electron mutual annihilation, for example, is

(C-1) $_1e^0 + {}_{-1}e^0 \Rightarrow \approx + \approx$ where $\approx$ is a photon of gamma radiation

It happens as follows [per equation *2-6*].

(C-2) $(_1e^0) + (_{-1}e^0) = U_c \cdot [1 - \cos(2\pi \cdot f_e \cdot t)] - U_c \cdot [1 - \cos(2\pi \cdot f_e \cdot t)]$
$$= 0$$

The two oscillations literally cancel. The annihilation occurs because the two are point-by-point inverses of each other. Such an annihilation is depicted in Figure C-1 on the following page.

In general for a particular particle and some particular anti-particle of it, their phases and frequencies will not be identical because of their different velocities and histories of relativistic frequency shifts. However, for them to mutually annihilate they must remain co-located for some brief moment sufficient for the event to occur.

For the particles to be co-located for a brief moment their positions and velocities must be identical, which means that their frequencies and their phases will also be identical.

The mutual annihilation energy is the conversion into energy of the entire mass of the two particles involved. The mass of each of the particles is its oscillation [there is nothing else to be the mass]. At annihilation the two particles' oscillations cease to exist by cancelling each other out. Since the center oscillations cease, the last waves of *Propagated Outward Flow* are followed by no U-waves at all from those centers.

Section *5 – Ampere's Law* showed that E-M radiation is the propagation of changes in the *Propagated Outward Flow*, changes usually caused by velocity changes of charged particles. The ceasing at annihilation of the oscillations of the two particles involved [the largest change possible] causes a pair of gamma photons, equation *C-1*, to be propagated.

The photons carry off conservation maintaining energy and momentum. The frequency of each photon is the frequency of the oscillation that just ceased, which corresponds to the mass of the particle. In other words the photon energy, $W = h \cdot f$, is the energy equivalent of the entire mass of the of the particle annihilated.

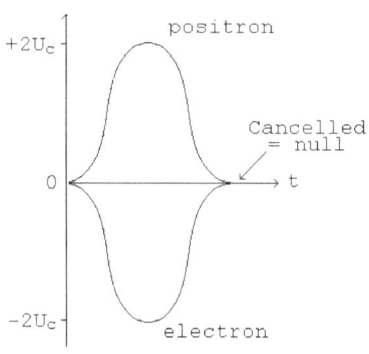

Figure C-1
A Mutual Annihilation

The first issue to investigate is the necessary conditions for a matter / antimatter annihilation to take place: how close must the particle and its antiparticle be and for how long must they remain in such sufficiently intimate contact ?

In addition to those two factors there is the more obvious requirement that the two particles involved be true antiparticles of each other [for example, a proton and an antiproton or an electron and a positron, but not a proton and a positron nor a proton and an electron]. Furthermore in general, particle / antiparticle annihilations are relatively unlikely between electrically neutral particles [for example, a neutron and an antineutron] because the only effects tending to bring the two together are their very weak gravitational attraction or chance encounter.

The Closeness Criterion

Indication of how close the two participating particles must be for their annihilation to take place can be found from the decay of a free neutron [not one that is part of an atomic nucleus] into a proton and an electron, a natural process with a mean lifetime before decay of about 881.5 seconds. For the neutron decay to be successful the proton and electron product particles must derive from the parent neutron not only their rest masses but also sufficient kinetic energy so that they are at escape velocity relative to each other, else they would be attracted back together and recombine. [One can neglect the also emitted electron anti-neutrino which is of zero or negligible mass.]

The escape velocity of the two particles is, at first consideration, an awkward problem because the separation distance of the two particles, which appears in the denominator of the expression for their Coulomb attraction, would seem to be required to be as small as zero. That is, at first consideration the escape velocity required is infinite. But, since infinite escape velocity is impossible yet the escape occurs, then the starting point, the minimum separation distance that can occur must be greater than zero. In other words, the neutron decay products, a proton and an electron, exist as such only when separated by some minimum Separation Distance, S, and their state at lesser separation distances appears as their parent neutron.

Therefore, since if the proton and the electron are separated by less than that minimum distance they do not exist as proton and electron but rather as the neutron, and at separation distances greater than that minimum they are the pair of separate particles, then that Separation Distance is a measure of how close a proton and an electron must be to unite into a neutron and is indicative of the spacing at which a particle and its antiparticle mutually annihilate.

The point is that the excess of the mass of the neutron over that of a proton plus that of an electron must supply the proton and electron relativistic kinetic masses needed to escape the decaying neutron. The detailed analysis and relativistic calculations can be found in Appendix A-1, *The Neutron.* The results are as follows.

(C-3) - The escape velocities:

$$v_e = 275,370,263. \quad \text{meters per second}$$
$$= 0.918,536,33 \cdot c$$
$$v_p = 379,350.6975 \quad \text{meters per second}$$
$$= 0.001,265,378 \cdot c$$

- The minimum Separation Distance:

$$S = 1.3 \cdot 10^{-15} \quad \text{meters}$$

Some years ago experiments involving measurement of the scattering of charged particles by atomic nuclei, yielded an empirical formula for the approximate value of the radius of an atomic nucleus to be

(C-4)   Radius $= [1.2 \cdot 10^{-15}] \cdot [\text{Atomic Mass Number}]$ meters

which formula would indicate that the radius of the proton as a Hydrogen nucleus (atomic mass number $A = 1$) is about $1.2 \cdot 10^{-15}$ meters.

The mass of the proton can be expressed as an equivalent energy, $W_p = m_p \cdot c^2$, and that as an equivalent frequency, $f_p = m_p \cdot c^2 / h$, or as an equivalent wavelength, $\lambda_p = c/f = h/m_p \cdot c$. That wavelength (not a "matter wavelength") for the proton is

(C-5) $\lambda_p = 1.321,410,0 \cdot 10^{-15}$ meters

quite near to the empirical value for the proton radius from equation *(C-4)* and the Separation Distance, *S*, of equation *(C-3)*. Thus the Separation Distance boundary between a proton and an electron as separate particles versus combined into a neutron is about *1* proton radius, the equivalent wavelength for the proton mass per equation *(C-3)*.

Then for a proton and an antiproton the boundary between their being the two separate particles and their mutually annihilating is a proton radius, a Separation Distance of $S_p = \lambda_p = 1.321,410,0 \cdot 10^{-15}$ *meters*. At that boundary if their velocities have a sufficient net component directly toward each other [per the time criterion, below] they would seem to be able, and likely, to mutually annihilate, and otherwise the annihilation would seem not possible.

Similarly, the mass of the electron or the positron can be expressed as the equivalent energy, $W_e = m_e \cdot c^2$, and that as its equivalent frequency, $f_e = m_e \cdot c^2 / h$, or equivalent wavelength, $\lambda_e = c/f = h/m_e \cdot c$. That wavelength (not a "matter wavelength") for the electron / positron is

(C-6) $\lambda_e = 2.426,310,6 \cdot 10^{-12}$ *meters*.

Then for an electron and a positron the boundary between their being the two separate particles and their mutually annihilating is a Separation Distance of $S_e = \lambda_e = 2.426,310,6 \cdot 10^{-12}$ *meters*. At that boundary if their velocities have a sufficient net component directly toward each other [per the time criterion, below] they would seem to be able, and likely, to mutually annihilate, and otherwise the annihilation would seem not possible.

Then, what is that sufficient net velocity ?

The Time Criterion

The mutual annihilation of a particle and its antiparticle is symbolized as in the following example for a proton and an antiproton.

(C-7) $_1p^1 + _{-1}p^1 \Rightarrow \gamma + \gamma$ where γ is a gamma photon

In the present case of a proton and an antiproton the mass of each of the protons is converted into the energy of the related γ photon. The frequency and period of each of those two photons is as follows.

(C-8) $f_{\gamma p} = m_p \cdot c^2 / h$

$T_{\gamma p} = 1/f_{\gamma p} = h/[m_p \cdot c^2] = 4.407,749,3 \cdot 10^{-24}$ *seconds*

In communications theory it is shown that a sinusoidal oscillatory signal must be sampled at least twice per cycle for the signal to be correctly represented. That is, two independent datums are required so as to determine the value of the oscillation's two absolute parameters, its amplitude and its frequency. [It's phase is relative, not absolute.] That implies that the time duration of a proton / antiproton mutual annihilation must be the period of each of the resulting photons.

(C-9) $\Delta t_{proton / antiproton} = T_{\gamma p} = 4.407,749,3 \cdot 10^{-24}$ *seconds*

Similarly for an electron / positron mutual annihilation, the time duration would be

152

(C-10) $\Delta t_{electron\ /\ positron} = T_{\gamma e} = 8.093,301,0 \cdot 10^{-21}$ seconds.

While those are very brief times they are not instantaneous.

In the case of a particle and its antiparticle coming together from significantly far apart, the particles will have accumulated significant velocity toward each other by the time they arrive at Separation Distance S because of having been accelerated by their mutual Coulomb attraction. However, the situation was different for the Big Bang.

WHY THE CRITERIA FAILED IN THE CASE OF THE BIG BANG

The number of particles resulting from the original Big Bang is estimated to have been about 10^{85} [Appendix B, *The Limitation of the Original Envelopes*], and those particles emerged on paths that were initially radially outward. The event was overall spherically symmetrical on the large scale, but at the local particle level perfect symmetry was impossible because of the nature of finite particles versus a smooth non-particulate substance. Initially all of the particles were on divergent paths although for two adjacent particles the amount of the divergence was minute.

For a proton and an adjacent antiproton in the Big Bang to be separate [not annihilated] at the instant of being projected outward in the Big Bang, they had to be separated by at least the above-developed $S_p = 1.321,410,0 \cdot 10^{-15}$ meters. For them to then annihilate their Coulomb attraction would have had to accelerate them into co-locating in the required time criterion starting from their initially zero velocity toward each other. [Actually they would have had non-zero but minute velocities away from each other because each follows its own outward radial path.] The issue is whether their Coulomb attraction can accelerate the two particles to the point of co-locating within the time frame of equation *C-9* [or equation *C-10* for an electron / positron case].

If, for example, for their mutual annihilation, the proton or the antiproton is to travel <u>at constant velocity</u> its half of the separation distance, $\frac{1}{2} \cdot S_p$, in time $T_{\gamma p}$, so as to be co-located with its antiparticle at the end of that time, it would require a speed of

(C-11)
$$v_p = \frac{\frac{1}{2} \cdot S_p}{T_{\gamma p}} = 0.5 \cdot c \qquad \text{[half light speed]}$$

and if the electron or the positron, for their mutual annihilation, is to travel its half of the separation distance, S_e, in time $\frac{1}{2} \cdot T_{\gamma e}$ <u>at constant velocity</u> it would require a speed of

(C-12)
$$v_e = \frac{\frac{1}{2} \cdot S_e}{T_{\gamma e}} = 0.5 \cdot c \qquad \text{[half light speed]}.$$

The achieving of that speed, if even only by the very end of the extremely short time period of the acceleration and travel, 10^{-21} *seconds or less*, would be difficult. The particles moving continuously at that <u>constant velocity</u> throughout their travel from separated to co-located is impossible in that they commence their travel of distance S from essentially zero velocity toward each other.

Furthermore, the analysis of the Coulomb interaction at close separation distances presented in Appendix A-1, *The Neutron* shows that the attraction weakens drastically at close quarters per Figure C-2, below, reproduced from that appendix. [The figure shows the form of the reduction in the Coulomb attraction as a function of the charge separation radial distance relative to a proton mass equivalent wavelength, λ_p.]

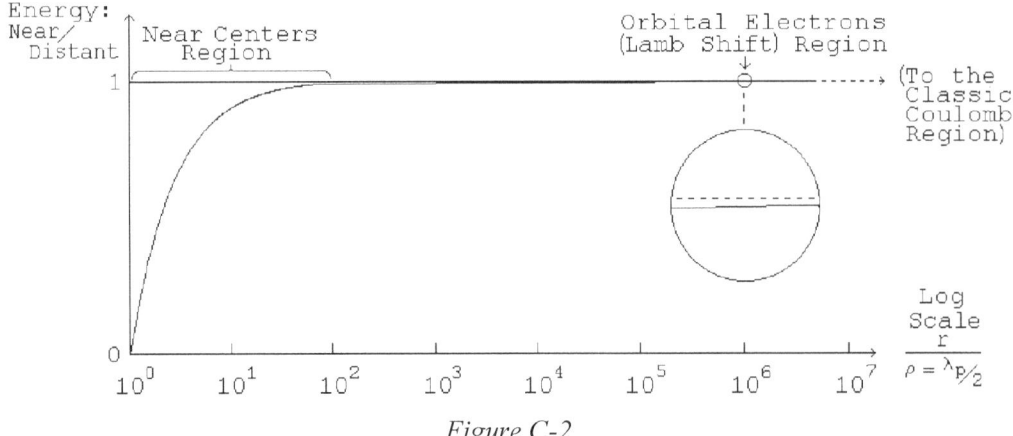

Figure C-2
Coulomb Effect <u>Reduction Factor</u> When Charges Are Near to Each Other

Finally, the posited particle and its antiparticle, emerging from the Big Bang, with spacing adjacent to each other as closely as possible, and on radially outward paths, were not alone. They were surrounded by a more or less uniform, symmetrical, large group of like particles and antiparticles. Any Coulomb tendency to unite the posited particle pair was largely offset by the similar tendency of each member to unite with the adjacent particle on its other side. The net Coulomb action on a specific particle or antiparticle was certainly insufficient to produce enough acceleration to enable the particle to transit its half of the Separation Distance in the required gamma photon period.

In summary:

- Adjacent Big Bang product particles and their antiparticles,

- Initially spaced optimally for co-locating [as closely as possible yet independently separate],

- Traveling outward at near light speed on essentially parallel paths [actually minutely diverging paths],

- Are unable to accelerate toward each other, from zero initial such velocity, quickly enough for their annihilation to produce the known actual gamma photons that would have to result from their mutual annihilation.

- That is, they cannot travel to the point of annihilation in time for the annihilation gamma photons to be the correct frequency to carry off the energy equivalent of the input particles, the pre-annihilation proton / antiproton or electron / positron.

In other words a Big Bang mutual annihilation was much more difficult, and rare, than one might have assumed. A large scale annihilation of matter and antimatter could not have taken place in the Big Bang. The result is that the present universe contains both matter and antimatter in equal amounts because of the original symmetry.

A UNIVERSE CONTAINING BOTH MATTER REGIONS AND ANTIMATTER REGIONS

Why Matter and AntiMatter Regions Are Able to Co-Exist

Of course, matter / antimatter mutual annihilations in general are not as awkward as they were for the original Big Bang with its peculiar initial conditions. Of interest here, however, is the case of the interstellar medium. It is the interstellar medium that must be examined because it is the natural boundary between regions of matter and regions of antimatter; where, if they are to occur, the anticipated matter / antimatter annihilations should be occurring and yielding their looked-for gamma ray flux.

In the interstellar [and intergalactic] medium the particles and antiparticles start from being significantly separated, residing in the vacuum of interstellar space, which vacuum, while not devoid of competing particles, has a much lower particle density than the original Big Bang. They do not suffer the disadvantage of being in a dense milieu of particles and antiparticles whose Coulomb attractions tend to cancel out their effects. And, they avoid the disadvantage of always starting their mutual Coulomb attraction toward each other with no initial velocity. Without regard for any mutual attraction between particular particles and antiparticles, they all move with significant velocities.

However, those velocities are in general not oriented toward the combination of a pair. Rather, the velocity directions are a combination of [a] some component distributed randomly over the particles in essentially all possible directions, and [b] some amount corresponding to a general flow direction.

Table C-1, below summarizes the particle [and antiparticle where applicable] content of interstellar space. The density of the particles, and their related mean distance apart are such as to militate against any significant number of encounters, whether aided by Coulomb attraction or not. [Excepting solar wind, which is local to star's nearby environment, most of the interstellar medium is Hydrogen atoms, not ions.] [Gravitation can be ignored here, it being decades of orders of magnitude weaker than Coulomb attraction.]

Region	Size	Particle	
		Density [/cc]	Energy
Our Solar Wind	Sun Neighborhood	10.	0.001- 0.004 × c
Our Local Cloud	60 Light Years	0.1	~ 7,000 °K
Our Local Bubble	300 Light Years	0.001	~ 1,000,000 °K
Intergalactic Space	[The Universe]	0.000 ... ?	?

Table C-1 – The Interstellar Medium

As has been pointed out in analyses of our solar wind, with typically *1 atom* in each *10 cm³* of interstellar gas in our local cloud and *10 ions* in each *cm³* of our solar wind, the particles are so far apart that the solar wind and interstellar gas flow through each other without being disturbed by collisions. On that basis, the even less dense regions of the interstellar medium such as ones like our local bubble, those within galaxies in general, and those in intergalactic space are even less conducive to particle / antiparticle encounters.

Another factor bearing on the likelihood of matter / antimatter mutual annihilations occurring in interstellar space is as follows. Because gravitational and Coulomb field attraction communicate at c, particles are attracted to where the attractor <u>was</u>, not where it <u>is</u>. That tends to produce orbital motion or "sling shot" non-collision passages rather than direct collisions. For example, a proton traveling at *0.000,001·c [only 300 meters/second]* and at a distance of *0.001 millimeter* from another charged particle [compare that distance with the spacing implied by the densities of the above table] will travel a distance equal to *757 of its proton radii* during the time that its Coulomb field communicates at velocity c to the other charged particle its then Coulomb attraction impulse.

All of these various factors taken into account, matter / antimatter collisions must be quite infrequent events in the interstellar medium. When such mutual annihilations occur the appropriate gamma photons are emitted.

Indications of Some Matter / AntiMatter Mutual Annihilations

A most likely indication of our detection of cosmic matter / antimatter annihilations is Gamma Ray Bursts [GRB's].

GRB's are flashes of gamma rays coming from seemingly random places in deep space at random times. GRB's last from milliseconds to minutes, and are often followed by "afterglow" emission at longer wavelengths. Gamma-ray bursts are detected by orbiting [*Swift*] satellites about two to three times a week. All known GRB's come from outside our own galaxy. Most GRB's come from billions of light years away [as much as *z = 6.3* or more].

Under the assumption that a given burst emits energy uniformly in all directions, some of the brightest bursts correspond to a total energy release of 10^{47} *joules*, nearly a solar mass converted into gamma-radiation in a small amount of time. No candidate process other than a significant matter-antimatter annihilation is able to liberate that much energy so quickly.

156

$\longrightarrow$

$\longrightarrow$

Appendix D

Integration Details for Magnetic Effect Calculations

PART (1) -- EQUATION 5 -14, THE STATIC CASE, ↑ COMPONENT

(5-14)
$$\uparrow F_E = \int_{-\infty}^{0} \uparrow dF(x) + \int_{0}^{+\infty} \uparrow dF(x)$$

$$= \int_{-\infty}^{0} F_r \cdot \frac{R^3}{[x^2 + R^2]^{1\frac{1}{2}}} \cdot dx + \int_{0}^{+\infty} F_r \cdot \frac{R^3}{[x^2 + R^2]^{1\frac{1}{2}}} \cdot dx$$

Since the form of the integral in each of the two regions is the same most of the integration process can be performed on just the form.

(D-1)
$$\uparrow \text{Form} = \int_{a}^{b} F_r \cdot \frac{R^3}{\left[x^2 + R^2\right]^{1\frac{1}{2}}} \cdot dx$$

$$= F_r \cdot R^3 \cdot \int_{a}^{b} \frac{1}{\left[x^2 + R^2\right]^{1\frac{1}{2}}} \cdot dx \qquad \text{[}F_r \text{ and R are constants for this integration]}$$

$$= F_r \cdot R^3 \cdot \frac{x}{R^2 \cdot \left[x^2 + R^2\right]^{\frac{1}{2}}} \Bigg|_{a}^{b} \qquad \text{[The integration anti-derivative]}$$

$$\qquad\qquad\qquad\qquad\qquad \text{[Divide through by } R^2 \cdot x \text{]}$$

$$= F_r \cdot R \cdot \frac{1}{\left[1 + x^2 \Big/ R^2\right]^{\frac{1}{2}}} \Bigg|_{a}^{b}$$

Returning to the overall equation 14-14 and evaluating at the limits:

(D-2)
$$\uparrow F_E = F_r \cdot R \cdot [0 - 1] \quad \text{For}: a = -\infty \quad R^2\!\big/_{\infty^2} = 0$$

$$\qquad\qquad\qquad\qquad b = 0 \quad R^2\!\big/_{0^2} = \infty$$

$$= F_r \cdot R \cdot [1 - 0] \quad \text{For}: a = 0 \quad R^2\!\big/_{0^2} = \infty$$

$$\qquad\qquad\qquad\qquad b = +\infty \quad R^2\!\big/_{\infty^2} = 0$$

$$= 2 \cdot F_r \cdot R \qquad\qquad \text{Overall}$$

159

PART (2) -- EQUATION 5 -30, CASES 1, 2, & 5, ↑ COMPONENT

(5-30)

$$\uparrow F_T = \int_{-\infty}^{0} \uparrow dF(v,x) + \int_{0}^{+\infty} \uparrow dF(v,x)$$

$$= \int_{-\infty}^{0} \uparrow f(v,x) \cdot dF(x) + \int_{0}^{+\infty} \uparrow f(v,x) \cdot dF(x)$$

$$= \int_{-\infty}^{0} \left[\left[\frac{A \cdot B \cdot (x^2 + R^2)}{\left[x^4 + (A^2 + B^2) \cdot R^2 \cdot x^2 + A^2 \cdot B^2 \cdot R^4 \right]^{\frac{1}{2}}} + \frac{(C+D) \cdot x}{\left[x^2 + R^2 \right]^{\frac{1}{2}}} \right] \left[F_r \cdot \frac{R^3}{\left[x^2 + R^2 \right]^{1\frac{1}{2}}} \right] \right] \cdot dx$$

$$+ \int_{0}^{\infty} \left[\text{The same above expression again for the second integration range} \right]$$

Integrating by parts the following is obtained for $\uparrow F_T$.

(D-3)

$$F_T = \int_{-\infty}^{0} f(v,x) \cdot dF(x) + \int_{0}^{+\infty} f(v,x) \cdot dF(x)$$

$$= \left[f(v,x) \cdot F(x) - \int F(x) \cdot df(v,x) \right]_{-\infty}^{0} + \left[f(v,x) \cdot F(x) - \int F(x) \cdot df(v,x) \right]_{0}^{+\infty}$$

for which f(v,x) and F(x) are as follows.

$$f(v,x) = \left[\frac{A \cdot B \cdot (x^2 + R^2)}{\left[x^4 + (A^2 + B^2) \cdot R^2 \cdot x^2 + A^2 \cdot B^2 \cdot R^4 \right]^{\frac{1}{2}}} + \frac{(C+D) \cdot x}{\left[x^2 + R^2 \right]^{\frac{1}{2}}} \right]$$

$$F(x) = F_r \cdot R \cdot \frac{x}{\left[x^2 + R^2 \right]^{\frac{1}{2}}}$$

The integration from $-\infty$ *to* 0 cancels out with the integration from 0 *to* $+\infty$ in the two integrals. The remaining portion of the integration is the $f(v,x) \cdot F(x)$ portion.

The value of $f(v,x)$ for $x=0$ is 1. Dividing the numerator and the denominator of the first term of $f(v,x)$ by x^2 and of the second term by x the value of the function for $x = \infty$ is $[A \cdot B + C + D]$. Similarly, $F(x)$ evaluates to $F_r \cdot R \cdot 1$ for $x = \infty$ and to $zero$ for $x = 0$.

Therefore the final result is

(D-4)
$$\uparrow F_T = F_r \cdot R \cdot [A \cdot B + C + D]_{>0} + F_r \cdot R \cdot [A \cdot B + C + D]_{<0}$$
$$= 2 \cdot F_r \cdot R \cdot [A \cdot B + C + D]$$

160

$\longrightarrow$

$\longrightarrow$

APPENDIX E

The Universal Exponential Decay

Since the "Big Bang" the *Propagated Outward Flow* has been gradually depleting the original supply of *medium* in the core of each *Spherical-Center-of-Oscillation*. That process, an original quantity gradually depleted by flow away of some of the remaining quantity, is an exponential decay.

THE NATURE OF THE DECAY

Of the three fundamental dimensions of length *[L]*, mass *[M]*, and time *[T]* only length can decay. Time being the independent variable of material reality, whether it decays, varies, or is rigorously constant is beyond our ability to detect. Likewise, mass cannot decay, it being proportional to frequency, the inverse of time.

The dimension that is decaying is length, the *[L]* dimension in the dimensions of, for example: the Planck Constant, h, *[M·L^2/T]*; the speed of light, c, *[L/T]*; and the Newtonian Gravitational Constant, G, *[L^3/$_M$·$_T2$]*. The decay process involves the fundamental constants (c, q, G, h, etc.) and decay of any of those must be dimensionally consistent with the decay of the others.

The *Propagated Outward Flow* is in Coulomb's Law the equivalent of q^2. Using the notation "{x}" to mean the dimensional units of "x", the dimensional units of *Propagated Outward Flow* i.e. *medium* are *[M·L]* per equation *(E-1)*.

(E-1)
$$\{Force\} = \{Mass\}\cdot\{Acceleration\} = M \cdot \frac{L}{T^2} = \{Coulomb\ Force\} = \frac{\{Q \cdot Q\}}{\{2\pi \cdot R^2\}} = \frac{\{Q^2\}}{L^2}$$

$$\{c \cdot q\} = \{Q\} = \frac{\sqrt{M \cdot L^3}}{T} = \frac{L}{T} \cdot \sqrt{M \cdot L}$$

$$\{q^2\} = M \cdot L = \{Propagated\ Outward\ Flow\}$$

The *Propagated Outward Flow* carries energy and momentum. Planck's constant, h, is the fundamental energy-related constant of physics. The ultimate reality that is within the *Spherical-Center-of-Oscillation*'s core is its supply of *medium* of dimensions *[M·L]*. To so be as h-based energy it is $h/_c$ which is of dimensions *[M·L]*.

The core's outer boundary is a surface of area $4 \cdot \pi \cdot \delta^2$. It lacks the power to restrain or contain anything. However, the only way the content of the core can leave and flow outward is through the core's surface. That flow is subject to the speed limit of light speed. That sets the flow at *[$4 \cdot \pi \cdot \delta^2$] · c*.

163

THE RATE OF THE DECAY

A process which the core decay resembles is the pumping of gas out of a chamber to create a vacuum. In this case the "gas" is the medium, the chamber is the core, and the pumping is the loss of medium, through the surface boundary of the core, to outward propagation. The process of the pumping, whether of gas out of a vacuum chamber or medium out of the core is such that:

· The rate of change of the amount of medium remaining in the core equals

· The amount per volume of medium remaining, times

· The pumping speed, that is the volume per time of the propagation.

This is based on the conceptualization of the process as:

- The medium is uniformly distributed in the core;

- A minute increment of volume is then pumped out in a minute time;

- The remaining medium then redistributes uniformly within the core,

- and the cycle repeats over and over.

From this the rate of change of the amount of medium present within the core is as follows.

(E-1)

$$\begin{bmatrix} \text{Medium} \\ \text{Rate of} \\ \text{Change} \end{bmatrix} = -\begin{bmatrix} \text{Amount} \\ \text{per} \\ \text{Volume} \end{bmatrix} \times \left(\begin{bmatrix} \text{Pumping} \\ \text{Speed} \end{bmatrix} = \begin{bmatrix} \text{Surface} \\ \text{of Core} \end{bmatrix} \times \begin{bmatrix} \text{Flow} \\ \text{Speed} \end{bmatrix} \right)$$

$$\frac{d\upsilon}{dt} = -\frac{\upsilon}{\frac{4}{3}\cdot\pi\cdot\delta^3} \times \left(\left[4\cdot\pi\cdot\delta^2 \right] \times [c] \right) = -\frac{3\cdot c}{\delta}\cdot\upsilon$$

The pumping takes place over the entire surface of the core and the rate at which the outward flow takes place is the speed of medium travel, the speed of light, c. [Both c and δ are functions of time, each decaying in its dimensional unit [L]; however, their decay rates are identical so that their ratio, as in equation (E-1) is constant.]

Therefore, rearranging equation (E-1) and integrating:

(E-2)

$$\frac{d\upsilon}{\upsilon} = -\frac{3\cdot c}{\delta}\cdot dt$$

$$\ln(\upsilon) = -\frac{3\cdot c}{\delta}t + C$$

$$\upsilon(t) = \upsilon_0\cdot\varepsilon^{-\frac{3\cdot c}{\delta}\cdot t} \qquad [\varepsilon^C \text{ evaluated as } \upsilon_0]$$

Therefore, the decay time constant, τ is

(E-3)

$$\tau = \frac{\delta}{3\cdot c}$$

However, that result cannot be correct. Equation (E-3) yields a value of about $4.5\cdot10^{-44}$ seconds. That is completely inconsistent with the universe having an already accomplished life time of billions of years.

164

It must be concluded that medium empties from the core at only a minute amount of the volumetric pumping speed used above or, alternatively, that the core volume contains, as medium, an immense supply of volume, of "highly concentrated volume" so to speak.

In Section 2, *The Behavior of Matter: Its Form,* under the subtitle "The Particle Core's Propagated Outward Flow", it was stated:

> "For such a flow to persist there must be a supply of that outward flowing substance in every particle. And, for that flow to have persisted the billions of years since the "Big Bang" that "supply" must be an <u>extremely concentrated reservoir</u> of that which flows outward ."

However thought of, it must be from the foregoing that an additional factor that reduces the rate of change of the core medium must be used in equation *(E-3)* so that it becomes

(E-4)

$$\begin{matrix}\text{Medium}\\\text{Rate of}\\\text{Change}\end{matrix} = -\begin{bmatrix}\text{Amount}\\\text{per}\\\text{Volume}\end{bmatrix}\times\left[\begin{bmatrix}\text{Pumping}\\\text{Speed}\end{bmatrix}=\begin{bmatrix}\text{Surface}\\\text{of Core}\end{bmatrix}\times\begin{bmatrix}\text{Flow}\\\text{Speed}\end{bmatrix}\right]\cdot\begin{bmatrix}\text{Concentration}\\\text{Factor}\end{bmatrix}$$

$$\frac{d\upsilon}{dt} = -\frac{\upsilon}{\frac{4}{3}\cdot\pi\cdot\delta^3}\times\left[\left[4\cdot\pi\cdot\delta^2\right]\times c\right]\cdot\frac{1}{F} = -\frac{3\cdot c}{\delta\cdot F}\cdot\upsilon$$

where *F* is the additional factor. Equations *(E-2)* and *(E-3)* then become *(E-5)* and *(E-6)* as follows.

(E-5)

$$\frac{d\upsilon}{\upsilon} = -\frac{3\cdot c}{\delta\cdot F}\cdot dt$$

$$\ln(\upsilon) = -\frac{3\cdot c}{\delta\cdot F}\cdot t + C$$

$$\upsilon(t) = \upsilon_0\cdot\varepsilon^{-\frac{3\cdot c}{\delta\cdot F}\cdot t}\qquad[\varepsilon^C \text{ evaluated as } \upsilon_0]$$

Therefore, the decay time constant, τ is

(E-6)

$$\tau = \frac{\delta\cdot F}{3\cdot c}$$

The values of δ and c are known, but what is the value of *F* ?

The gradually decaying medium contained within the core is not merely the geometric core physical volume as viewed from our world; it is "highly concentrated volume", the capability if freed into space outside the core to be myriad core physical volumes, the volume of space.

That difference distinguishes the physics of the core's internal "Core Domain" vs. the outside "World Domain".

The ratio to the world view geometrical volume of that *"highly concentrated volume"* <u>of medium to be propagated</u> is designated *F.*

(E-7)
$$F = \frac{\text{Volume Equivalent of Core Medium Supply}}{\text{Geometric Core Volume}}$$

$$= \frac{h/c}{4/3 \cdot \pi \cdot \delta^3} \; \frac{\text{Units } [M \cdot L]}{\text{Units } [L^3]} = 7{.}938{,}010{,}000 \cdot 10^{60}$$

F is a pure number just as are $4/3$ and π of equation *(E-7)*. Saying the core is medium $[M \cdot L]$ vs. volume $[L^3]$ is like saying a year is [days] vs. [seconds].

The factor F spans two different regimes of material reality:

1 - The natural world regime in which we exist and function;

2 - The interior of the core of each particle, the supply of highly concentrated medium, minute portions of which are propagated outward in each cycle of the particle's oscillation, gradually depleting the supply.

The factor F spans the relationship between the "Core" and "World Domains"; it expresses the connection of the physical volume of the core and the concentrated-volume medium filling the core. It converts expressing the interior of the core, its substance, between units of volume, $[4/3 \cdot \pi \cdot \delta^3]$ { Length^3 }, and units of medium $[h/c]$ { Mass $\cdot$ Length }, as propagated outward.

From equation (E-6) with the value for F of equation (E-7) the value of τ, the universal decay time constant is

(E-8) $\quad \tau = 3.57532 \cdot 10^{17}$ seconds

$\quad\quad\quad \approx 11.3373 \cdot 10^{9}$ years

[See the scientific article *On Five Independent Phenomena Sharing a Common Cause,* available at the physics archive at www.arXiv.org , paper arXiv:physics/0101003 [pdf] .

The paper describes a succession of five different independent astronomical phenomena each the result of a common underlying cause that produces an unaccounted-for acceleration that is: quite small, centrally directed in the system exhibiting each phenomenon, non-gravitational, distance independent, and of a common magnitude.

It demonstrates that the underlying common cause of the five otherwise unexplained phenomena is the Universal Exponential Decay.]

$\longrightarrow$

$\longrightarrow$

BOOK 2 – GRAVITATIONAL APPLICATIONS

CONTENTS

$\longrightarrow$

SECTION 8

Deflecting "Propagated Outward Flow"

INTRODUCTION

Because gravitation operates by a <u>flow</u> from the attract<u>ing</u> mass acting on the attract<u>ed</u> mass as shown in Section 7, it is possible by deflecting that flow to partially deflect gravitation away from an object so that the gravitational attraction on the object is reduced. That effect makes it possible to extract energy from the gravitational field, which makes the generation of *gravito-electric* power technologically feasible. Such plants would be similar to hydro-electric plants and would have the hydro-electric advantages of not needing fuel and not polluting the environment but, they are much less expensive and can be located anywhere, not needing special sites for dams.

Physically, the action of deflecting gravitational attraction, which of course is directed <u>toward</u> the gravitation source, produces an equal but opposite reaction on the deflecting mechanism directed <u>away</u> from the gravitation source. The result is the combination of reducing the gravitational attractive acceleration of the object toward the gravitation source plus the introducing of a reactive acceleration on the object away from the gravitation source.

For example, a deflector experiencing a natural gravitational acceleration, A, reduced by 100% gravitation deflection to $0.5 \cdot A$, plus simultaneously experiencing the reaction to the 100% deflection in the amount $1.0 \cdot A$, experiences a net acceleration acting in the direction <u>away</u> from the gravitation source of $1.0 \cdot A - 0.5 \cdot A = 0.5 \cdot A$. Of course, the A is the Newtonian gravitational acceleration $G \cdot M / d^2$ where M and d are the mass of and distance to the gravitating source, for example the Sun, the Earth, or Mars.

Such a deflector, engineered with controlled adjustment of the amount and the direction of its action, could provide spacecraft launch levitation and deep space travel acceleration. It could provide both levitation and horizontal motion for a flying vehicle over a planet surface.

Just as the sail-driven ships of past centuries experienced fuel-free travel by means of controlling the energy of the wind, this technology enables fuel-free travel through space by controlled manipulation of the gravitational field that permeates all of space. It uses readily abundantly available materials and techniques and is ready now for research and engineering refinement.

171

THE PROPAGATED OUTWARD FLOW FROM ALL MATTER

The result of Sections 1 through 7 is that all matter is *Spherical-Centers-of-Oscillation* from which *Propagated Outward Flow* is continually taking place and which through its various interactions produces all of the physical effects that we experience, light and gravitation in particular for the present subject.

As pointed out in Section 2, when the original oscillation came into existence it did so in absolute nothing. The μ_0 and ε_0 of the *Propagated Outward Flow* were then part of that flow the only thing they could have come from and are so now. Each particle's outward *Propagated Outward Flow* contains its own μ_0 and ε_0.

That *Propagated Outward Flow* is an oscillatory wave of *[1 - cosine]* form. It carries and produces the effects of gravitation by its effect on propagation speed determined by the parameters μ and ε. It carries and produces the effects of light as a modulation of its oscillatory waveform. Both are carried by the same one *Propagated Outward Flow* oscillatory wave.

For *Spherical-Centers-of-Oscillation*, which propagate waves of oscillating flow the factors that determine the flow's speed of propagation are [like those of a transmission line] the time required to build up the flow amount for each oscillation cycle through each successive infinitesimal increment of the flow's μ_0 and to build up the flow's potential for each oscillation cycle on each successive infinitesimal increment of the flow's ε_0. The speed, radially outward, of its propagation, is c per equation *8-1*.

$$(8\text{-}1) \qquad \text{Speed} = c = \frac{1}{\sqrt{\mu_0 \cdot \varepsilon_0}}$$

Because the *Propagated Outward Flow* is radially outward its concentration is reduced inversely as the square of the distance from the source *Spherical-Center-of-Oscillation*. That progressively reduces the concentration of μ and ε in the flowing medium but it likewise reduces the magnitude of the flow and therefore the time required to build it up through each infinitesimal increment of the flow's μ and similarly for the flow's ε. As a result the inverse square dispersion does not affect the speed of flow.

Upon encountering another particle's flow the **a**rriving flow's μ_a and ε_a combine with the μ_e and ε_e in the **e**ncountered flow the *$[\mu_a + \mu_e]$* and *$[\varepsilon_a + \varepsilon_e]$* sums being larger values for the overall μ and ε. The result is that the encountered flow is slowed relative to its natural speed, equation *8-1*, and likewise the arriving flow.

That process changes the speed of flow because the amount of μ and ε is a scalar quantity so that the arriving and encountered quantities combine when co-located. But, the flow through each infinitesimal increment of the μ and the potential on each infinitesimal increment of the ε are vector quantities. Their time for vector build up through each infinitesimal increment of the combined flow's μ and ε are for the arriving and the encountered flows separate independent processes of each even though the scalar μ's and ε's combine and each flow must address that combination.

Unless an external event deflects it, each vector flow increment along its *Propagated Outward Flow* wave front periphery pursues its own radial outward propagation direction hedged-in by its flow increments next immediately on each side

172

while taking longer to build up flow and potential increments in the common greater μ and ε as $[\mu_a + \mu_e]$ and $[\varepsilon_a + \varepsilon_e]$.

Therefore *Propagated Outward Flow* from a *Spherical-Center-of-Oscillation* encountering another *Spherical-Center-of-Oscillation* and its *Propagated Outward Flow* operates to slow the propagation speed of both the flows. One body's gravitational flow can affect another body's gravitational flow. One body can change another body's gravitational affect on other body's.

DEFLECTION OF PROPAGATED FLOW

Light normally travels in a straight direction. But, when some effect slows a portion of the light wave front the direction of the light is deflected. In Figure 8-1 below, the shaded area propagates the arriving light at a slower velocity, v', than the original velocity, v, so that the direction of the wave front is deflected from its original direction.

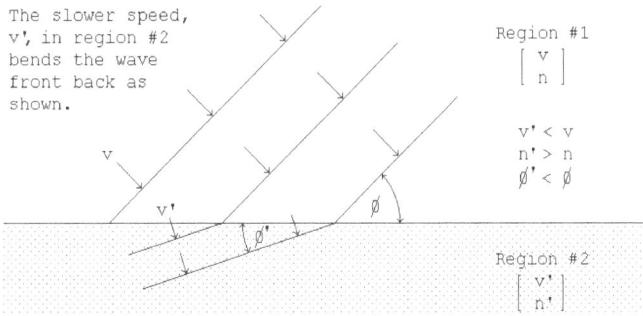

Figure 8-1
Deflection of Light's Direction by Slowing of Part of Its Wave Front

"Gravitational Lensing", Figure 8-2, is another example of slowing of part of a wave front resulting in curving or deflecting the direction of flow.

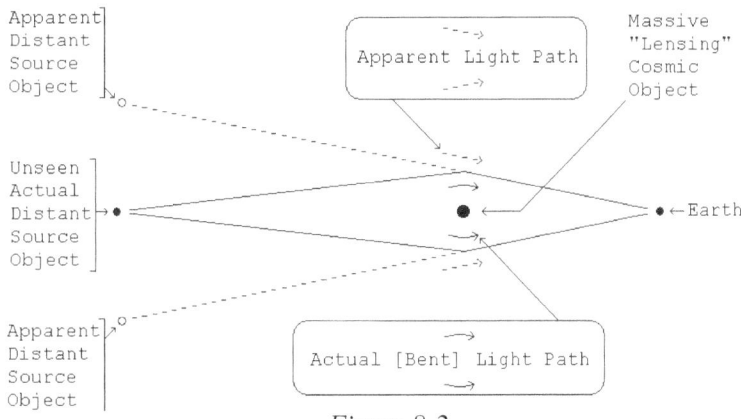

Figure 8-2
Gravitational Lensing Bending of Light Rays

"Gravitational lensing" is an astronomically observed effect in which light from a cosmic object too far distant to be directly observed from Earth becomes observable because a large cosmic mass [the "lens"], located between the Earth observers and that distant object, deflects the light from the distant object as if focusing it, somewhat concentrating its light toward Earth enough for it to be observed from Earth.

The light rays are so bent because the inverse square reduction in the lensing object's *Propagated Outward Flow* slows more the portion of the incoming light's *Propagated Outward Flow* wave front that is nearer to the lens than it slows the farther away portion of the light wave front.

The same effect occurs on a much smaller scale in the diffraction of light at the two edges of a slit cut in a flat thin piece of opaque material as shown in Figure 8-3 below. The bending is greater near the edges of the slit because the slowing is greater there. The effect of the denser material in which the slit is cut slows the portion of the wave front that is nearer to it more than the portion of the wave front in the middle of the slit where there is only air.

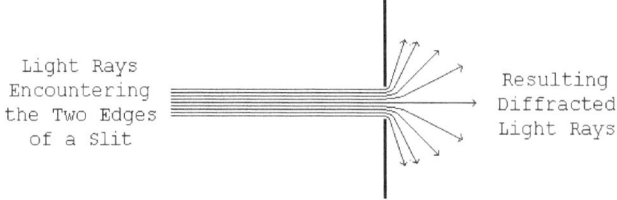

Figure 8-3
Diffraction at a Slit Causing Bending of Light Rays

In both of these cases, gravitational lensing and slit diffraction, the direction of the incoming *Propagated Outward Flow* wave front is changed because part of the wave front is slowed relative to the rest of it. In the case of gravitational lensing the part of the wave front nearer to the "massive lensing cosmic object" is slowed more. In the case of diffraction at a slit the part of the wave front nearer to the solid, opaque material in which the slit is cut is slowed more. Both effects are because each body's slowing-causing *Propagated Outward Flow* is reduced as the square of the distance from its source.

In the slit diffraction effect the role of the "massive lensing cosmic object" is performed by the individual atoms making up the opaque thin material in which the slit is cut. That shows that the gravitational lensing process, involving immense cosmic masses, can be implemented on Earth on a much smaller scale practical for human use.

Therefore, a properly configured material structure can deflect gravitation away from its natural action by deflecting the gravitation's *Propagated Outward Flow,* reducing the natural gravitation effect on objects that the gravitation would otherwise encounter and attract.

THE ENERGY ASPECT AND THE SOURCE OF THE FLOW

But, changing the "natural gravitation effect" means changing the gravitational potential energy of objects in the changed gravitational field. If the energy is changed where does the difference come from or go to ?

The potential energy for an object of mass, m, at a height, h, in a gravitational field is truly <u>potential</u>. It is the kinetic energy that the mass <u>would acquire</u> from being accelerated in the gravitational field <u>if it were to fall</u>.

While at rest at height h [as on a shelf] the total mass of the object is the same as its rest mass. The object has no actual "potential energy". It is merely in a situation

174

where it could acquire energy by falling in the gravitational field. Falling, the mass of the object increases as its velocity increases, reflecting its gradually acquired kinetic energy.

Since, <u>until it falls, the object does not have the energy that it will acquire when it falls</u> in the gravitational field <u>the energy that it acquires must come from the gravitational field during the fall</u>.

The energy of gravitational field is in its *Propagated Outward Flow* radially outward from all gravitational masses. The *Propagated Outward Flow* is a flow of the potential for energy, realized at any encounter with another mass

- That *Flow* creates potential energy, <u>creates the situation where kinetic energy could be acquired</u>, at any mass that it encounters.
- It does so continuously, replenished and replenishing by the on going continuous *Propagated Outward Flow*.
- It does so continuously, regardless of the number of masses or amount of mass that the *Propagated Outward Flow* encounters and regardless of their distance from the source of the *Flow*.

But, for there to be a continuous *Flow* outward from each mass particle, each must be a supply, a reservoir, of that medium which is flowing. That reservoir supplying the on-going continuous *Propagated Outward Flow* is presented in Book I, Section 2. The original supply of the *Flow* medium, of gravitational potential energy, came into existence at the beginning of the universe.

If that immense reservoir of energy could be tapped by tapping some of its appearance in its outward *Flow*, which is the gravitational field, it could be a vast supply of energy cheaply, cleanly, and permanently without [for practical human / Earth purposes] being used up.

[Since the "Big Bang" the *Propagated Outward Flow* has been gradually depleting the original supply. That process is an exponential decay. The time constant is about $\tau = 3.57532 \cdot 10^{17} \; sec \; (\approx 11.3373 \cdot 10^9 \; years)$].

Tapping the Energy of the Gravitational Field

The general vertically upward *Propagated Outward Flow* of gravitational energy can be tapped by deflecting part of a local region's gravitational *Flow* away from its normal vertical direction. Figure 8-4 below (the slit diffraction of Figure 8-3 now rotated 90°) illustrates such deflection using a single slit.

Resulting Deflected Rays of
Flow of Gravitation

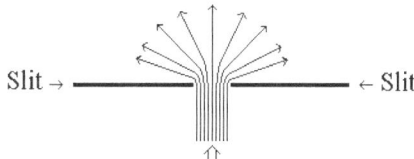

Slit → ———————— ← Slit

Rays of Flow of Gravitation
Encountering the two Edges of a Slit

Figure 8-4
Slit Diffraction, the Basic Element of a Gravitation Deflector

175

Multiple such slits parallel to each other would spread the deflection left and right in the figure. Additional multiple such slits at right angles to the first ones would spread the deflection over a significant area.

GRAVITATION DEFLECTOR DESIGN CONCEPT

The edges of the slit in the above Figure 8-4 are actually rows of atoms. A cubic crystal, such as of Silicon, consists of such rows of atoms, multiple straight line rows and rows at right angles, all equally spaced – a naturally occurring configuration of the set of slits required for deflection of gravitation. (The atoms are at the corners of the cubes; the lines, not natural, indicate the cubic structure.)

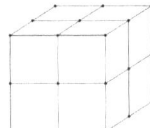

Figure 8-5
A Small Piece of a Cubic Crystal

The *Flow* from each of the cubic crystal's atoms is radially outward. Therefore its concentration falls off as the square of distance from the atom. The amount of slowing of an incoming gravitational *Flow,* and therefore the amount of its resulting deflection, depends on the relative concentrations of the atoms' *Flow* and the overall gravitational *Flow.*

In the case of diffraction of the *Flow* of light at a slit the concentration of the *Flow* from the atoms of the slit material is comparable to the concentration in the horizontal *Flow* of the light, because it originates from a local source, not from the Earth's immense gravitation.

But for the *Flow* from the atoms of the slit to deflect the much more concentrated vertically upward *Flow* of Earth's gravitation the *Flow* from the atoms of the slit must also be much more concentrated. The only way to achieve that more concentrated *Flow* is a configuration in which the *Flow* of Earth's gravitation is forced to pass much closer to the atoms of the slit so that, per the inverse square variation in the atoms' *Propagated Outward Flow*, the *Flow* of Earth's gravitation will pass through a concentration of the slit atom's *Flow* comparable to the concentration in the Earth's gravitational *Flow.*

The spacing between the edges of the diffracting slit [Figure 8-4] is about $5 \cdot 10^{-6}$ *meters*. The spacing of the atoms at the corners of the "cubes" in a Silicon cubic crystal is $5.4 \cdot 10^{-10}$ *meters*. An inter-atomic spacing of less than $3 \cdot 10^{-19}$ *meters*, much closer than the natural spacing in the Silicon cubic crystal, is required to obtain deflection of a major portion of the incoming Earth's gravitational *Flow,* per Appendix E, Relative *Propagated Outward Flow* Concentrations.

Such a close atomic spacing cannot be obtained by directly arranging for, or finding a material that has, such a close atomic spacing. However, that close an atomic spacing can be effectively produced relative to just the vertical *Flow* of gravitation by slightly tilting the Silicon cubic crystal's cubic structure relative to the vertical.

Figure 8-6 on the following page illustrates the tilting, schematically not to scale, and shows how it increases the number of crystal atoms closely encountered by the upward gravitational *Flow.*

By appropriate tilting of the cubic structure each of its $5.4 \cdot 10^{-10}$ *meters* inter-atomic spacing is effectively sub-divided into 10^{10} "sub-spaces" each of them $5.4 \cdot 10^{-20}$ *meters* long and with an atom in each. A *4.5 mm* shim on a *30 cm* diameter Silicon cubic crystal ingot produces such an effect, producing a tilt *tangent = 0.015* for a *tilt angle = 0.86°* that produces the objective effective sub-division of the crystals' natural inter-atomic spacing, a sub-division that acts only on vertical *Flow*, as of gravitation.

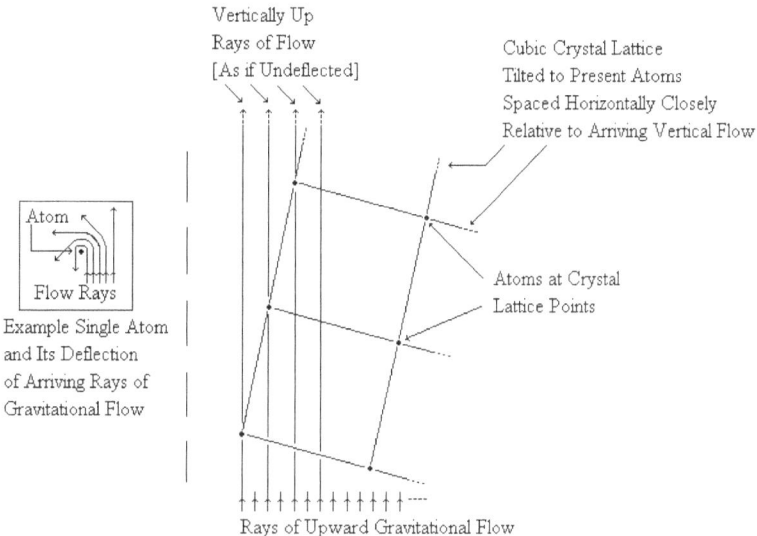

Figure 8-6
Cubic Crystal Lattice Tilted for Effective Gravitational Flow Deflection

Pure, monolithic, Silicon cubic crystals up to *30 cm* in diameter are grown for making the "chips" used in many electronic devices. The gravitation deflector requires a large, thick piece of Silicon cubic crystal rather than the thin wafers sawed from the "mother" crystal for "chip" making.

Mean free path *[MFP]* is the average straight line distance a moving particle travels between encounters with another particle. For atoms in solid matter the mean free path is

$$(8-2) \quad \text{MFP} = \frac{1}{[\text{Atoms Per Unit Volume}] \cdot [\text{Atom Cross Section Area}]}$$

For the Earth the atoms per unit volume is on the order of

    Atoms per Unit Volume = $5 \cdot 10^{28}$ per cubic meter.

In the cubic crystal deflector the atomic spacing produced by the tilt is about 10^{-20} *meters*. Each therefore has cross sectional space available to it of that of a circle of that diameter so that for this purpose the atom's cross section area is

    Atom Cross Section Area = $\pi/4 \cdot [10^{-20}]^2$
                            = $8 \cdot 10^{-39}$ meter^2

For targets as fine as those in the cubic crystal deflector, the mean free path in the Earth's outer layers is, therefore

    MFP = $2.5 \cdot 10^9$ meters

177

The mean free path in the thick minutely tilted Silicon cubic crystal ingot for intercepting Earth's natural <u>vertically</u> outward gravitation is $\frac{1}{2}$ the thickness of the ingot. For a *50 cm = 0.5* m thick ingot the gravitation deflector is about 10^{10} times more effective than the natural Earth at intercepting Earth's natural gravitation. However, that effectiveness is only for vertical rays of *Flow*. The Silicon crystal's mean free path for non-vertical *Flow – Flow* already once deflected within the crystal – is that of Earth, $2.5 \cdot 10^9$ *meters*, which takes the once-deflected *Flow* out of the crystal.

$\longrightarrow$

$\longrightarrow$

Quantifying the Deflection

The manner of the deflection of the *Propagated Outward Flow* is curving of the path of rays of the flow as they pass close to atoms of the deflector with the direction to which curved depending on the relative positions of the ray and an atom and the amount of the curving depending on how close the ray passes to the atom. Because of the range of those variables and their various combinations the "deflection" is essentially a "scattering" in various amounts in various directions, all scattering being away from the perfectly vertical upward which the deflector is designed to solely deflect by virtue of its atomic spacing and slight tilt.

A two-dimensional physical example of the deflection is the diffraction pattern of light diffracted by a slit. Figure 9-1, below, presents the diffraction pattern produced by a slit that is $5.4 \cdot 10^{-6}$ *meter* wide with incoming light of wavelength $4.13 \cdot 10^{-7}$ *meter*. The peaks and valleys of the pattern, the interference pattern, are a phenomenon of the light imprint on the *flow* that carries it. The envelope of the pattern is the relative amounts of the underlying *flow* carrying the light.

For that reason, while the interference pattern varies according to the wavelength of the light involved, the form of the envelope of that pattern is always the same.

Diffraction Pattern

·Slit = $5.4 \cdot 10^{-6}$ Meter Wide
·Light Wavelength = $4.13 \cdot 10^{-7}$ Meter

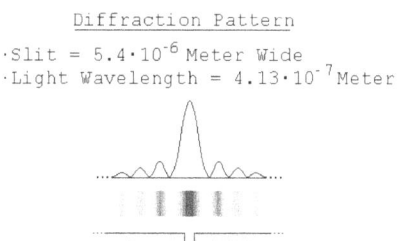

Figure 9-1
A Slit Light Diffraction Pattern

The *flow* concentration produced by the two slit edges falls off with distance from the edge inversely as the square of distance from its atoms. The Cauchy-Lorentz Distribution is an inverse square function of its variable. Its Density Function can represent the relative *flow* intensity pattern produced by the diffraction process by representing the envelope of the diffraction pattern. In Figure 9-2, below, the Cauchy-Lorentz distribution is fitted to the diffraction pattern by the appropriate choice of value of its distribution parameter γ [Greek *gamma*].

<u>The Envelope of the Relative Intensities of the</u>
<u>Light Diffraction Pattern Is the Actual Amount</u>
<u>of the *Flow* Relative Intensities</u>.

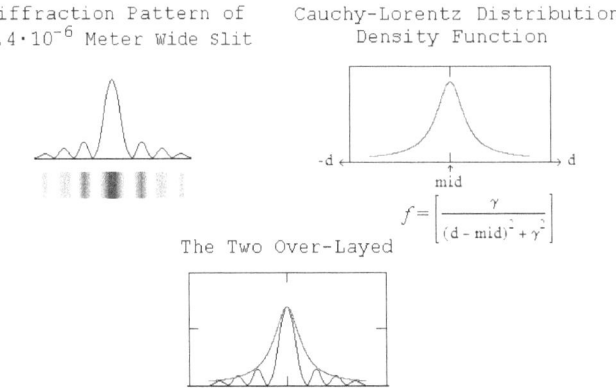

Diffraction Pattern of   Cauchy-Lorentz Distribution
5.4·10⁻⁶ Meter Wide Slit   Density Function

The Two Over-Layed

Figure 9-2
The Cauchy-Lorentz Distribution Diffraction Pattern Envelope

The deflection angle, Φ, is the angle of deflection of the rays to any particular point on the diffraction pattern. That is Φ is the angle of deflection of the rays directed to that particular point and of intensity per the Cauchy-Lorentz Distribution at that point.

The interest here is not in the location of the light interference maxima and minima, but in the deflection angles the diffraction imposes on the *flow*. However, calculation of the deflection angles to the minima provides a good indication of the amount of *flow* deflection obtained over the overall diffraction pattern. The table below presents that data for the $5.4 \cdot 10^{-6}$ *meter* wide slit with incoming light of wavelength $4.13 \cdot 10^{-7}$ *meter*. [The minimums are counted outward from the center peak of the diffraction interference pattern].

Minimum #	$\Phi°$		Minimum #	$\Phi°$
1	4.39		8	37.72
2	8.80		9	43.50
3	13.26		10	49.89
4	17.81		11	57.28
5	22.48		12	66.60
6	27.36		13	83.86
7	32.37		14	$Sin(\Phi) > 1.0$

Sin(Φ) = n · [light wavelength / slit width], n = 1, 2, ...
Figure 9-3 – Table of Diffraction Minimums Deflection Angles

Again, while we are not interested in the diffraction minimums and not in the diffraction interference patterns at all, the envelope of the diffraction pattern depicts the distribution of the deflection of the *flow* that carried the light in the diffraction pattern.

The above table demonstrates that the deflection of the *flow* is at least in amounts up to $90°$. That deflection may well extend to angles beyond $90°$, perhaps to as much as $180°$, a complete reversal of direction. There is no way of determining that from the diffraction pattern, however, because the light of the diffraction pattern cannot be deflected beyond $90°$ in any case because the light cannot penetrate the material containing the slit.

But, the *flow* readily penetrates and permeates all of material reality.

The tilt [Figure 1-6] of the cubic crystal structure divides the slit into 10^{10} sub regions the first and last of which are at the slit's edge and produce the maximum deflection. The tilt also arranges that ultimately all of the vertical components of the incoming vertical flow must pass through one of those "at the edge of the slit" regions, must experience maximum deflection.

The overall average effect is equivalent to every ray's vertical component curving at least $90°$ because the crystal tilt causes every ray to pass extremely close to an atom at some point in the crystal, as shown for the extreme rays in the figure below.

Resulting Deflected Rays of
Flow of Gravitation

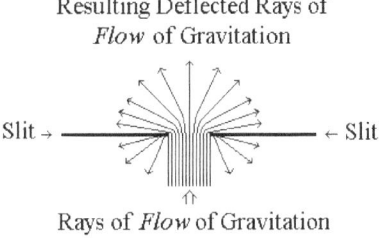

Slit →   ← Slit

Rays of *Flow* of Gravitation
Encountering the two Edges of a Slit

Figure 9-4 – Single Slit Gravitation Deflection

PROPAGATED OUTWARD FLOW DEFLECTION CAUSED BY WAVE SLOWING

The bending of Propagated Outward flows' paths results from differential slowing, that is the systematic slowing of the flow wave front in different amounts along that front. The slowing takes place in accordance with equation $1-1$, above. Figure 9-1, below, depicts the differential slowing-caused process.

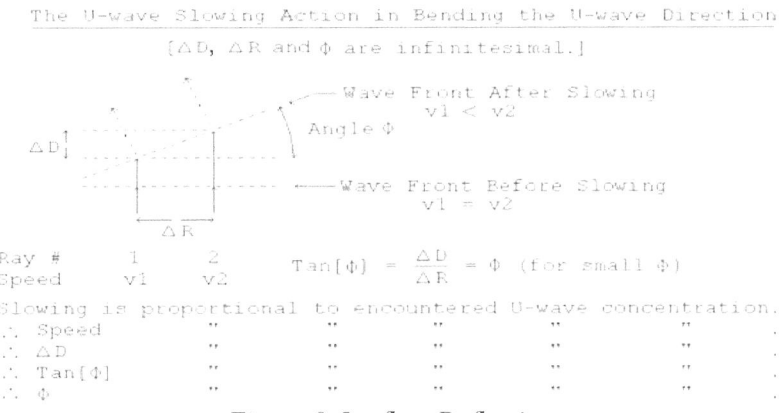

Figure 9-5 – flow Deflection

The figure indicates the differential slowing of the upward-directed [as for gravitation] flow flux that results in deflection of the flows' paths. The slowing is directly proportional to the encountered concentration of the encountered flow flux, and, therefore the angle of deflection, Φ, is proportional to that concentration.

QUANTIFYING THE FLOW DEFLECTION IN LIGHT DIFFRACTION

The diffraction pattern is a projection on a screen or piece of photographic film of the diffracted light as it spreads out due to the diffracting action. The physical size, the linear dimension of the pattern becomes larger as the distance from the diffracting slit to the screen or film on which the pattern appears increases. But the angles, as measured from the center of the slit to any point on the diffraction pattern [relative to the 0° angle from the center of the slit to the center of the pattern], are the same regardless of the distance from the slit to the screen or film.

Therefore, to analyze and evaluate the pattern requires attending to those angles, not linear distances on the pattern. Since the linear distances on the pattern are irrelevant, any convenient distance from the slit to the screen or film may be chosen. In the following analysis that distance will be taken as equal to the slit width, $5.4 \cdot 10^{-6}$ meter in this case.

The data of interest here, which is a measure of the amount of flow bending contained in the diffraction pattern, is the portion of the total light incident on the slit appearing in any specified portion of the diffraction pattern. That portion can be defined in terms of the angles just described and that portion is an otherwise dimensionless number, again independent of the physical or linear size of the diffraction pattern.

The Cauchy-Lorentz Distribution for this application is as follows.

(9-1) The Cauchy-Lorentz Distribution Density Function

[a] <u>In General</u>

$$f(x;x_0,\gamma) = \frac{1}{\pi} \cdot \left[\frac{\gamma}{(x - x_0)^2 + \gamma^2}\right]$$

[b] <u>As Used Here</u>

$$f(d;mid,\gamma) = \left[\frac{\gamma}{(d - mid)^2 + \gamma^2}\right]$$

mid = half-way point between slit edges
d = distance from mid
γ = half-width at half-maximum

From the above Figure 9-2, the half-width of the Cauchy-Lorentz Distribution at its half-maximum is 74.0% of the distance from the mid-point to the first minimum in the interference pattern. That is γ is 74.0% of the displacement from the centerline to the first intensity minimum outward from the centerline. Calculating the deflection angle to that minimum[4] the angle is found to be $4.39°$.

The corresponding displacement along the d-axis [for screen distance = slit width] of the above Figure 9-3 is the value of γ in the formulation of the Cauchy-Lorentz distribution.

(9-2) γ = [74% of] [[slit width]·Tan[4.39°]]

$\qquad$ = [0.74]·[5.4·10^{-6} meter]·[0.077]

$\qquad$ = 3.1·10^{-7} meter

The deflection angle, Φ, for any particular point on the diffraction pattern is the angle between [a] a reference line that runs from the center of the slit perpendicular to the barrier containing the slit toward the projected diffraction pattern and [b] a line running from the center of the slit to the location of the particular point on the diffraction pattern. That is the angle of deflection of the rays directed to that point and of intensity per the Cauchy-Lorentz Distribution at that point.

In these diffraction patterns so long as the ratio of the wavelength of the incident light to the width of the slit is constant, then each deflection angle, Φ, is independent of the distance from the slit to the screen where the diffraction pattern is projected.

The Cauchy-Lorentz Distribution's Cumulative Distribution Function is the integral of the Density Function, that is the area under the Density Function curve, the cumulative density. That function is given in equation *9-3*, below.

(9-3) The Cauchy-Lorentz Distribution Cumulative
$\qquad$ Distribution Function

$\qquad$ [a] In General

$$f_{cum}(x;x_0,\gamma) = \frac{1}{\pi}\cdot\arctan\left[\frac{x-x_0}{\gamma}\right] + \frac{1}{2}$$

$\qquad$ [b] As Used Here

$$f_{cum}(d;mid,\gamma) = \frac{1}{\pi}\cdot\arctan\left[\frac{d-mid}{\gamma}\right] + \frac{1}{2}$$

With $mid = 0$, when $d = -\infty$ [a deflection of $90°$ to the left in Figure 9-3], then $f_{cum} = 0$. Likewise at $d = +\infty$ then $f_{cum} = 1$, the total amount. To find the fraction, F, of the total amount of the incident light entering the slit that is deflected through some chosen angle, Φ, or more to the left of mid the procedure is as follows, taking $\Phi = -45°$ as an example and using $\gamma = 3.1\cdot10^{-7}$ meter per equation *9-2*. Because that light exists only on the flows carrying it the portion, F, is the fraction of the total amount of flows entering the slit that are deflected through angle Φ or more.

$\qquad$ 1 – Calculate the displacement, d, of Figure 9-3.

(9-4) d = Tan[θ] × [slit width]

$\qquad$ = Tan[-45°] × [5.4·10^{-6}]

$\qquad$ = -5.4·10^{-6} [for this example of θ = -45°]

$\qquad$ 2 – Calculate $F = f_{cum}(d;mid,\gamma)$ from equation *9-3*.

(9-5)

$$F = f_{cum}(d;mid,\gamma) = \frac{1}{\pi}\cdot\arctan\left[\frac{d-mid}{\gamma}\right] + \frac{1}{2}$$

$$= \frac{1}{\pi}\cdot\arctan\left[\frac{(-5.4\cdot10^{-6}) - (0)}{3.1\cdot10^{-7}}\right] + \frac{1}{2}$$

$$= 0.018$$

Then P, the percentage deflected through angle Φ or more of the total flows incident on the slit is:

$$F \div f_{cum} (d = +\infty) = F \div 1 = F.$$

P = 1.8% of total incident light entering the
     slit on each side [for this example].

In this example calculation the portion of the total flow flux that is deflected by $\Phi = 45°\ or\ more$ is $P_{45} = 1.8 + 1.8 = 3.6\%$.

Table 9-6, below, presents the portion of the total amount of the incoming gravitational flow flux that is deflected through some chosen angle, Φ or more, using the above $45°$ example type of calculations for each of the deflection angles cited in Table 9-3, above.

$\Phi°$	% Deflected	$\Phi°$	% Deflected
4.39	40.9	37.72	4.7
8.80	22.6	43.50	3.8
13.26	15.2	49.89	3.1
17.81	11.3	57.28	2.3
22.48	8.8	66.60	1.6
27.36	7.1	83.86	0.4
32.37	5.7		

Table 9-6
Percent of Total flow that is Deflected By Various Angles of Deflection, Φ, or More

USING THESE SLIT DIFFRACTION RESULTS FOR A GRAVITATION DEFLECTOR

The above table and example indicate that significant flow ray deflection does take place in the case of the atoms along the edge of the $5.4 \cdot 10^{-6}\ meter$ wide slit, but the amount of deflection is not very much – about only 3.6% deflected $45°$ or more, in the example.

On the other hand, looking at 100% of the rays of flow flux that arrive, uniformly spaced, at the $5.4 \cdot 10^{-6}\ meter$ wide slit, 3.6% of them arrived at that slit near enough to the atoms of one of the edges so as to be deflected $45°$ or more. All of the rays of that 3.6% achieved that much deflection because they passed their deflecting atom much more closely than the rest of the rays.

The 1.8% on each side of the Cauchy-Lorentz Distribution passed its deflecting atom within a distance of 1.8% of the slit width $[0.018 \times (5.4 \cdot 10^{-6}) = 9.7 \cdot 10^{-8}\ meter]$. If it could be arranged that all of the vertically upward flow gravitational flux were to pass that closely to atom then 100% of the gravitational flux would be deflected by $45°$ or more.

However, these deflection calculations are for a flow flux of the density or concentration of the flow carrying the beam of light to the diffracting slit. The vertically upward flow flux of the Earth's gravitational field is immensely more dense or concentrated

$\longrightarrow$

$\longrightarrow$

Cubic Crystal Deflector Calculations

A CUBIC CRYSTAL DEFLECTOR

A small portion of a Silicon cubic crystal is depicted below.

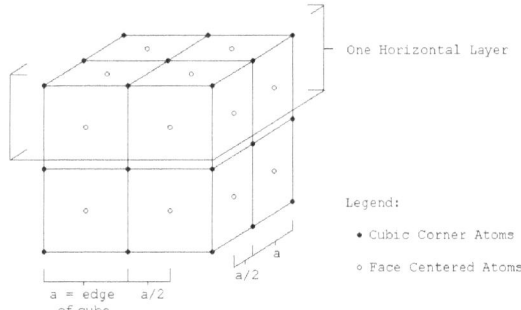

Figure 10-4
A Silicon Cubic Crystal

In the Silicon cubic crystal the edge of the cube, a, is $5.4 \cdot 10^{-10}$ *meters*. The effective horizontal interatomic spacing for vertically upward traveling *Flow* is half the edge, $a/2 = 2.7 \cdot 10^{-10}$ *meters*. From the figure the vertical layer thickness is $a = 5.4 \cdot 10^{-10}$.

The edges of the slit that produces the light diffraction pattern of Figures 9-1 and 9-2 on which the Section 9 analysis of Flow deflection is based consist of atoms spaced along the slit edge at an interatomic spacing that is essentially the same as a cubic crystal's interatomic spacing, about $2.7 \cdot 10^{-10}$ *meter*. The only difference between the light diffraction $5.4 \cdot 10^{-6}$ *meter* wide slit and a cubic crystal's interatomic spacing is that in the cubic crystal the "slit" width is that same interatomic spacing, about $2.7 \cdot 10^{-10}$ *meter*.

The diffraction pattern of Figures 9-1 and 9-2 is determined by the edges of the slit. The edges are the limit of the "slice" of incident light that passes through the slit and the light at those edges is the most deflected because it is the nearest to the deflecting atoms of the slit edge. Similarly, the edges jointly define the mid point of the diffraction pattern which is where the action of the two edges are equally strong so that their deflecting effects cancel each other to no net deflection.

In the cubic crystal those defining points are only apart $2.7 \cdot 10^{-10}$ *meter* as compared to $5.4 \cdot 10^{-6}$ *meter* apart in the case of the slit. The calculations of equations 9-1 through 9-5 must be re-calculated for that $2.7 \cdot 10^{-10}$ *meter* slit. That requires evaluating γ for its Cauchy-Lorentz Distribution. That is the same as in equation 9-2 except that the value of the slit width is changed to $2.7 \cdot 10^{-10}$ *meter*. The result is equation *9-2'*.

(9-2') γ' = [74% of] [[slit width]·Tan[4.39°]]

$\qquad$ = [0.74]·[$2.7\cdot10^{-10}$ *meter*]·[0.077]

$\qquad$ = $1.5\cdot10^{-11}$ meter

$\qquad$ Calculating the portion, P, of the total amount of the incident *Flow* entering that slit that is deflected through θ = -45° to the left of the mid point of the diffraction pattern and its Cauchy-Lorentz Distribution using γ = $1.5\cdot10^{-11}$ meter per equation *9-2'* is as follows.

$\qquad$ 1 – Calculate the displacement, d, of Figure 9-3.

(9-4') $\quad$ d = Tan[θ] × [slit width]

$\qquad$ = Tan[-45°] × [$2.7\cdot10^{-10}$]

$\qquad$ = $-2.7\cdot10^{-10}$ $\quad$ [this example of θ = -45°]

$\qquad$ 2 – Calculate P = $f_{cum}(d;mid,\gamma)$ from equation 9-3.

(9-5')
$$P = f_{cum}(d;mid,\gamma) = \frac{1}{\pi}\cdot\arctan\left[\frac{d - mid}{\gamma}\right] + \frac{1}{2}$$

$$= \frac{1}{\pi}\cdot\arctan\left[\frac{(-2.7\cdot10^{-10}) - (0)}{1.5\cdot10^{-11}}\right] + \frac{1}{2}$$

$$= 0.018$$

$\qquad$ Again the portion of the total Flow flux that is deflected by θ = 45° or more is P_{45} = *1.8%* + *1.8%* = *3.6%*. The result is unchanged from that in the case of the $5.4\cdot10^{-6}$ *meter* slit. The reason for that is that the parameters of the Cauchy-Lorentz Distribution describing the deflected Flow amounts in the various directions of deflection are determined by the two opposed slit edges. Contracting their spacing correspondingly contracts the distribution.

$\qquad$ Now for the *1.8%* on each side of the Cauchy-Lorentz Distribution to pass its deflecting atom within a distance equal to *1.8%* of the slit width, the new value of that distance is the value for the cubic crystal slit, *[0.018 × (2.7·10⁻¹⁰)* = *4.9·10⁻¹² meter]*. If it could be arranged that all of the vertically upward Flow gravitational flux were to pass within that close a distance of an atom of the cubic crystal lattice, then *100%* of the gravitational flux should be deflected by *45°* or more.

EARTH'S GRAVITATION VS. A SURFACE LIGHT SOURCE

$\qquad$ However the light-and-slit analysis deflections and calculations in <u>Section 3</u> were for light traveling in the Flow flux density generated by the Earth surface light source not the much more concentrated Earth overall gravitational Flow outward flux. The deflections and calculations for diffraction of light as developed in <u>Section 3</u> must be adjusted to compete at the level of Earth gravitational Flow flux rather than at that of an Earth surface light source if there is to be a noticeable deflecting affect on Earth gravitation.

$\qquad$ Appendix A is a detailed calculation of the relative gravitational strengths of natural objects at the Earth's surface and the Earth's surface overall planetary gravitation. The ratio of the Earth's surface gravitational acceleration, $9.8 \, m/_{sec2}$, to, from Table A-8 of Appendix A, the gravitational acceleration of air, $4.81 \times 10^{-17} \, m/_{sec2}$, is about $2 \cdot 10^{17}$. From that table, the gravitational acceleration of metals is on the order of $10^{-14} \, m/_{sec2}$ as compared to the Earth's overall gravitational acceleration of about $9.8 \, m/_{sec2}$ for a ratio of about 10^{15}. Consequently, the flux actually carrying the light [generated by a metallic light source] and entering the slit is the dominant factor not ambient air.

The Flow fluxes are proportional to the acceleration that they produce. The ratio of the accelerations, which is the ratio of the Flow fluxes, is as given in equation 10-1, below.

$$(10\text{-}1) \quad \text{Ratio} = \frac{\text{Acceleration of Earth Gravity}}{\text{Acceleration of Diffracted Light Flows}}$$

$$= \frac{\text{Earth Gravity Flows Flux}}{\text{Slit Diffracted Light Flows Flux}}$$

$$\approx 10^{15}$$

Therefore the Flow concentration which all vertical rays of gravitational Flow flux must be forced to encounter by being forced to pass close to the cubic crystal's atoms must for this purpose be made 10^{15} times greater. The gravitational Flow flux must be forced to pass accordingly even closer to the cubic crystal's atoms.

However, the Flow concentration from the atoms is inverse-square reduced with distance from the atom and accordingly so increases with nearness to the atom. Consequently, to increase the concentration by a factor of 10^{15} requires reducing the separation distance by a factor of only the square root of that, about $3.2 \cdot 10^7$.

The earlier above found effective interatomic spacing to be forced by tilting the cubic crystal, $2 \times [4.9 \cdot 10^{-12}] = 9.8 \cdot 10^{-12} \, meter$, must now be that divided by $3.2 \cdot 10^7$ the result for which is $3 \cdot 10^{-19} \, meter$. That arrangement, arranging that all of the Flow gravitational flux must, at some layer, pass within $3 \cdot 10^{-19} \, meter$ of an atom of the cubic crystal will result in essentially 100% of the gravitational Flow flux passing so close to some atom that it should be deflected by $45°$ or more..

With the cubic crystal's natural interatomic spacing being $2.7 \cdot 10^{-10} \, meter$ and the effective spacing to be forced is $3 \cdot 10^{-19} \, meter$ then each natural interatomic space must be sub-divided into $9 \cdot 10^8$ "pieces". If the crystal is tilted such that each of the layers of the crystal lattice is located offset from the layer below it by $[1/_{9} \cdot 10^8] \cdot [2.7 \cdot 10^{-10}] = 3 \cdot 10^{-19} \, meter$ in each of the two horizontal directions of the orientation of the lattice then the objective is met.

The direct implementation of that would require a tilt at an angle whose tangent is the offset divided by the interatomic [layer-to-layer] spacing, $[3 \cdot 10^{-19}] \div [5.4 \cdot 10^{-10}] = 5.6 \cdot 10^{-10}$, an angle of about $6.4 \cdot 10^{-8}°$. That means that the tilt causes each successive layer to offer its atoms a further $3 \cdot 10^{-19} \, meter$ offset so that enough layers will produce offering the atoms at every $3 \cdot 10^{-19} \, meter$ increment in each $2.7 \cdot 10^{-10} \, meter$ horizontal interatomic space. See Figure 10-5, next page.

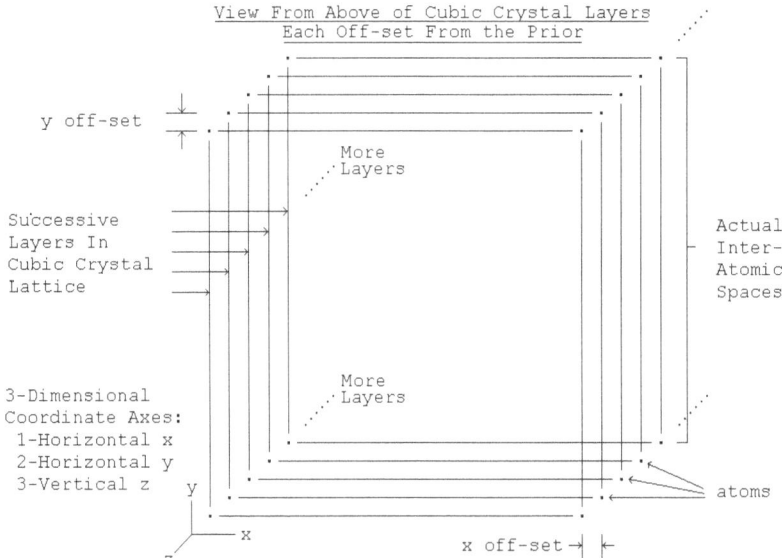

Figure 10-5
Crystal Layers, Offset Slightly, Achieving Effective
Close Interatomic Spacing
[The z-axis is vertical. The x- and y-axes are in layers.]
[Not to Scale.]

The required number of layers is one layer for each of the $9 \cdot 10^8$ "pieces" into which each $2.7 \cdot 10^{-10}$ *meter* horizontal interatomic space is divided: $9 \cdot 10^8$ layers.

Such a fine tilt angle and its precision are unlikely if not impossible to set up. The solution to that is that the successive layers need not each supply the minute offset relative to their adjacent layers. If the layers as depicted in Figure 10-5, above, were shuffled into any order whatsoever, they would still have the same effect that no vertical ray could avoid passing within $3 \cdot 10^{-19}$ *meter* of an atom, some atom, not necessarily one in the immediately next layer.

Of course, the layers in the cubic crystal cannot be shuffled or re-arranged, but that is not necessary. All that is necessary to operate using a larger tilt angle is that the same sufficient number of layers overall be employed and that the tilt be such ["workable tilt"] that the actual *x-axis offset* and the actual *y-axis offset* be such that, after that "same sufficient number of layers overall", each required effective atomic position appears somewhere, in some layer, even though not necessarily in "sequential order".

"Unworkable tilts" are those that duplicate needed atomic positions or that fail to produce all needed atomic positions, or both. The problem of what successfully workable tilts are and how they relate to unworkable tilts is developed in Appendix B, Factors Affecting Crystal Tilt.

The number of layers required, $9 \cdot 10^8$, requires a cubic crystal thickness of that number of layers multiplied by the individual layer thickness, which is $9 \cdot 10^8 \times 5.4 \cdot 10^{-10} = 0.50$ *meters* or *50 cm.*

Each "slit" in the cubic crystal is a pair of atoms spaced apart horizontally by the crystal lattice interatomic spacing of $2.7 \cdot 10^{-10}$ *meter*; or, more precisely, each "slit" is a linear "string" of such atom pairs, in any single layer of the crystal, and running from one side to the other of the crystal just as the slit edges in the case of light diffraction by a slit is a linear "string" of the atoms of which the slit edge consists.

For each such $2.7 \cdot 10^{-10}$ *meter* wide "slit" the above tilt procedure implements arranging that out of all of that portion of the total gravitational Flow flux that passes through it, the $1/_{9 \cdot 10^8}$ or $1.1 \cdot 10^{-7}$% on either side, a total of $2.2 \cdot 10^{-7}$%, passes within $3 \cdot 10^{-19}$ *meter* of an atom of the cubic crystal lattice and will be deflected by $45°$ or more away from its pre-deflection vertically upward direction. That leaves the issue of what happens to the balance of the gravitational Flow flux entering each such "slit".

The Flow propagation of each such atom falls off in concentration inversely as the square of the distance from it. The $3 \cdot 10^{-19}$ *meter* closeness is required to obtain the $45°$ deflection. Of the total gravitational Flow flux entering that slit, at ten times farther away from an atom, $3 \cdot 10^{-18}$ *meter*, the concentration is reduced by a factor of $[1/_{10}]^2 = 1/_{100}$. There the angle of deflection is reduced by approximately that factor to about $0.45°$. That deflection is experienced by about $1.1 \cdot 10^{-6}$% on either side out of the total gravitational Flow flux that passes through the "slit", a total of $2.2 \cdot 10^{-6}$%.

Still farther away, at $3 \cdot 10^{-17}$ *meter* from an atom, the concentration is reduced by a factor of $[1/_{100}]^2 = 1/_{10,000}$. There the angle of deflection is reduced by approximately that factor to about $0.0045°$ and applies to about $2.2 \cdot 10^{-5}$%.

Thus far more than 99 % of the total Flow flux entering the "slit" experiences negligible deflection. That is, until layer-by-layer in the crystal lattice further portions having earlier experienced that negligible deflection then experience the "$3 \cdot 10^{-19}$ *meter*" condition until, eventually, all of the Flow flux experiences the "$3 \cdot 10^{-19}$ *meter*" condition and is deflected by $45°$ or more.

Returning to the mean free path analysis at equation *1-2* it is now found to be the case that the target interatomic spacing to be achieved by the tilt of the cubic crystal is $3 \cdot 10^{-19}$ *meters* instead of 10^{-18} *meters*. The mean free path in the Earth for that same $3 \cdot 10^{-19}$ *meters* target size then is calculated as follows.

$(10-2)$ $\quad \text{MFP} = 1/_{C \cdot A}$

$$= \frac{1}{[\text{C, Atoms Per Unit Volume}] \times [\text{A, Atom Cross Section Area}]}$$

For the Earth the concentration of atoms is on the order of $C = 5 \cdot 10^{28}$ *per cubic meter.* In the cubic crystal deflector the target spacing achieved by the tilt is $3 \cdot 10^{-19}$ meters. Each target has cross sectional area space available to it equal to a circle of that diameter so that

$(10-3)$ $\quad \text{A} = \pi/_4 \cdot [3 \cdot 10^{-19}]^2 = 7.1 \cdot 10^{-38}$ *meter*^{2}

193

and, for such targets the mean free path per equation *1-2* in the Earth's outer layers is

(10-4) $\text{MFP} = 2.8 \cdot 10^8 \text{ meters.}$

That is to be compared to the mean free path in the cubic crystal deflector being one-half the cubic crystal thickness of *0.50 meters* or *0.25 meters*.

The gravitation deflector is about 10^{10} times more effective than the natural Earth at intercepting Earth's natural gravitation.

However, that effectiveness is only for vertical rays of *Flow*.

The Silicon crystal's mean free path for non-vertical *Flow – Flow* already once deflected within the crystal – is that of Earth, *2.5·10⁹ meters*, which takes the once-deflected *Flow* out of the crystal.

$\longrightarrow$

$\longrightarrow$

SECTION 11

Deflector Design Details

GENERAL DEFLECTOR DESIGN

In general the deflector consists of the following.

- A support having a verified perfectly horizontal upper surface for the cubic crystal deflector bottom face to rest upon;

- The Silicon cubic crystal ingot for the deflector as follows:
    - *30 cm* in diameter,
    - *50 cm* or more thick,
    - with the orientation of the cubic structure marked for placement of the tilt-generating shims, and
    - with the bottom face of the cylinder sawed and polished flat at a single cubic structure plane of atoms.

- Precision shims *4.5 mm* thick for producing the tilt of the cubic crystal ingot, the shims located at the mid-point of two adjacent sides of the horizontal plane of the cubic structure as in Figure 11-7 below.

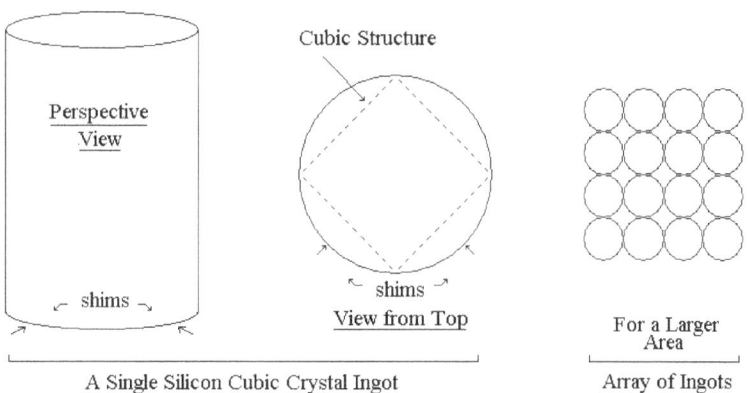

Figure 11-1
The Silicon Cubic Crystals Arrangements

- For an array of ingots for a larger area than a single ingot can provide, the individual ingots can be machined to fit snugly together. That could be done by machining them to a square cross section or, better, to a hexagonal one.

197

PRACTICAL ASPECTS AND ENGINEERING

   While the net gravitational field is vertically upward, i.e. radially outward from the Earth's surface, local gravitation is radially outward from each particle of matter. As in Figure 11-8 below, a mass above the Earth's surface receives rays of gravitational attraction from all over its surrounding surface and the underlying body of the Earth.

   The net effect of all of the rays' horizontal components is their cancellation to zero however the effect of all of the rays' vertical components is Earth-radially-outward gravitation.

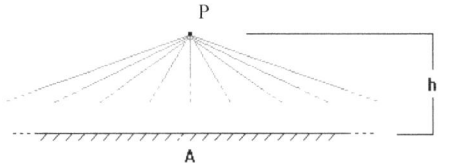

   < Earth Surface

Figure 11-2
Rays of Gravitation from the Surroundings

1 - *Gravitational Ray's Horizontal and Vertical Components.*

   One can consider all of the net gravitational effect on objects as being due to the vertical component of all of the myriad rays of gravitational field *Flow* at a wide variety of angles to the horizontal.

   The various rays of the *Flow* propagation from the individual particles of the gravitating body [e.g. Earth] are from each individual particle of it to the selected point [above the gravitating body] on which their action is being evaluated. That is the point *P* in the above Figure 11-8.

   The Earth's gravitational action along a ray of *Flow* takes place from the Earth's surface to deep within the Earth. The inverse square effect, that the strength of a *Flow* source is reduced as the square of the increase in the radial distance of it from the object acted upon, is exactly offset by that the number of such sources acting [per "ray" so to speak] increases as the square of that same radial distance. That is, the volume, hence the number, of *Flow* sources for a ray of propagation at the object is contained in a conical volume, symmetrically around the ray with its apex at the object acted upon.

   However, because the net gravitational effect is produced only by the vertical component of each ray of *Flow* propagation, the effectiveness of each ray is proportional to the Cosine of the angle between that ray and the vertical angle θ in Figure 11-9 below.

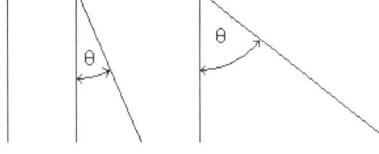

Figure 11-3
The Gravitational Field Ray Angle to the Vertical

   The actual total gravitational action includes all rays from $\theta = 0$ through to $\theta = 90^\circ$. That range would require an infinitely large deflector to act on all such rays. That is the deflector would have to be a disk of infinite radius. For lesser values of the maximum θ addressed, the portion of the total gravitation sources included is the integral of $Cos\ \theta \cdot d\theta$ from $\theta = 0\ to\ \theta = Chosen\ Lesser\ Value$. The integral of the *cosine* is the *sine*. Example lesser portions of the total gravitational action addressed as θ varies are presented in the table below.

θ	Sin θ = Fraction of T Gravitational Action
0°	0.000
30°	0.500
45°	0.707
60°	0.866

The gravitational deflector as a disk beneath the *Object* to be levitated must extend horizontally far enough to intercept and deflect the $Chosen\ Lesser\ Value$ of angle θ rays of gravitational wave *Flow* that are able to act on the *Object* of the deflection as depicted in Figure 11-10 below.

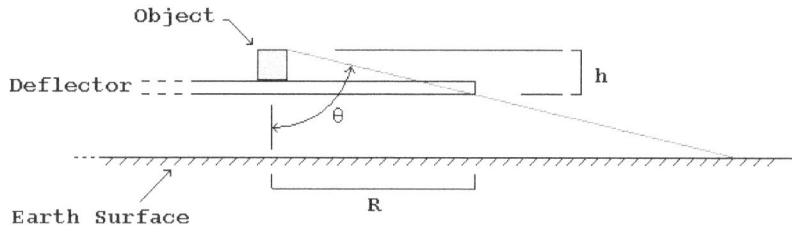

Figure 11-4
Size Requirements for a Disk Shaped Deflector

For the perfectly vertically traveling rays of gravitation waves the required vertical distance that must be traveled within the cubic crystal is the previously presented $50\ cm$ and 0 horizontal distance is traversed in so doing. But a ray at angle θ, in order to traverse the required 50 cm vertically, must traverse horizontally $50 \cdot Tan[\theta]\ cm$, at the same time. For θ more than 45° that and the deflector can become quite large.

Because the deflector disk must extend over a large area to deflect most of the gravitation, an alternative, and better, solution to the problem of rays of gravitation arriving over the range from $\theta = 0\ to\ \theta = 90^{\circ}$ is to wrap the deflector up the sides of the *Object* to be levitated as shown below.

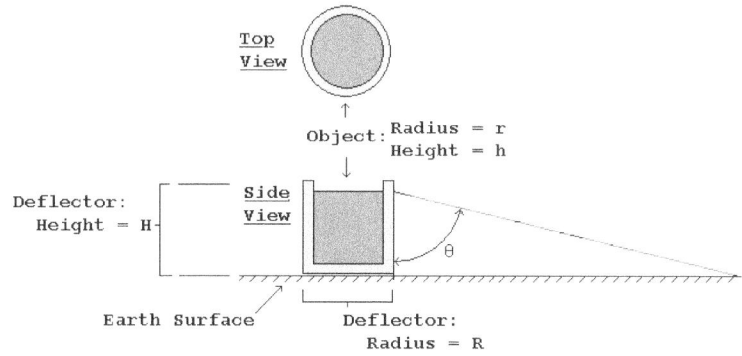

Figure 11-5
A Cup Shaped Gravitation Deflector

199

In this configuration the deflector takes up little more space than the Object levitated. However, the non-perfectly vertical traveling rays must still travel within the cubic crystal the horizontal distance $50 \cdot Tan[\theta]$ *cm*. That requires that the horizontal thickness of the vertical sides of the cup-shaped deflector must be of that $50 \cdot Tan[\theta]$ *cm* thickness.

Because the value of $Sin \; \theta$ and, therefore, the fraction of the total gravitational action, increases relatively little above $\theta = 60^{\circ}$ whereas the value of $Tan[\theta]$ increases quite rapidly, from $1.7 \; to \; \infty$ above $\theta = 60^{\circ}$ that $\theta = 60^{\circ}$ is the appropriate value to which to design. The thickness of the "walls" of the "cup" would then be $50 \cdot Tan[60^{\circ}] = 85$ *cm*. The deflector would be only slightly larger than the *Object* levitated.

2 - *The Array Structure and Size.*

The Deflector consists of an array of Silicon cubic crystals. The crystals forming the disk-shaped "base" of the "cup" need to be $0.5 \; m$ in height to achieve their maximum deflection effectiveness. Those forming the "sides" of the cup can be the same kind of $0.5 \; m$ crystals stacked and aligned vertically.

The crystals can effectively be grown in diameters up to about $30 \; cm$, however those cylindrical pieces must then be machined down to hexagonal cross section so that a number of them can fit together with negligible open space between. The hexagonal cross section area would be about $A = 0.06 \; m^2$

For an *Object* to be acted upon by the deflector, the object of height, h, and diameter, d, *meters* the deflector would have the following parameters for $\theta = 60^{\circ}$. [The number of crystals must be the integer next higher than the calculated number.]

Base Disk: Thickness = 1 Crystal Layer = 0.5 m
              Diameter   = d
              Area   = $\pi \cdot d^2 / 4$ = $0.785 \cdot d^2$
              Number of crystals = $\pi \cdot d^2 / 4 \cdot A$
                             = $13.1 \cdot d^2$

Cup Sides:

    Thickness   = 0.85 m
    Outside diameter [OD] = d + 2 · thickness
                      = d + 1.7
    Inside diameter [ID] = d
    Height   = h + 2 · 0.5
                      = h + 1.0

    Height number of Layers = $Height / 0.5$

    Area of Layer = $\pi \cdot [OD^2 - ID^2] / 4$

    Layer Number of crystals = $\pi \cdot [OD^2 - ID^2] / 4 \cdot A$

Total Number of Crystals:

    Number of Crystals =

        = Base Disk + [Layer Number $\times$ Number of Layers]

Some examples of these data are presented in the table below.

d	h	Cup Disk Base		Cup Sides			Total Crystals
		Area	Crystals	Nr of Layers	Area	Crystals	
1	1	0.785	14	2	4.94	99	212
10	10	78.5	1,310	20	28.97	580	12910

3 - *Calibrating the Individual Silicon Crystals*

The individual crystals making up the deflector cannot be grown exactly identical to each other. In each the orientation of the long axis of the cubic crystal structure may vary minutely from each of the others. That is, it is not certain that each crystal's base is purely a single plane of atoms of the cubic structure and thus is exactly perpendicular to the long axis of the crystal.

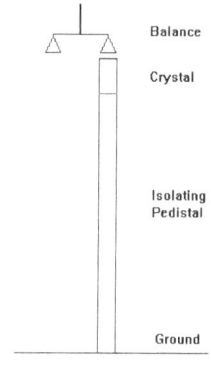

To find the optimum tilt and orientation for a single crystal the tilt must be varied over the range of possibilities while the effect of gravitation from exactly below it is observed on a balance scale. But most of the effect of gravitation on a single crystal is not from exactly below it.

The solution to that problem is to conduct the optimization atop a structure, that relying on the inverse square effect, effectively isolates the crystal from most of the gravitation from surrounding sources except that exactly below it – a high pedestal having a cross section comparable to that of the crystal, as in Figure 11-12.

Figure 11-12

To conduct that calibration on thousands of crystals should not be necessary if a method can be developed to exactly measure the long axis orientation in any given crystal. The process can then determine the optimum orientation of the crystal tilt relative to the actual long axis of a few cubic crystals being calibrated. That same crystal tilt relative to the actual long axis can then be applied to each of the other crystals.

The long axis orientation problem could also be solved by insuring that the base of each crystal is a single plane of atoms of the cubic structure.

Monolithic silicon cubic crystals are commercially available with the ends nearly a single plane, that is within *0.2 degrees* of the *(100)* plane of the cubic structure. In view of the various effects analyzed in Appendix B, and their resolution in its section *The Random Distribution Solution to The Crystal Tilt*, that amount or moderately more of inaccuracy in the crystal tilt may be of no significance except that it potentially may call for crystal thicknesses moderately greater than *0.5 m*.

201

$\longrightarrow$

SECTION 12

Gravito-Electric Power Generation

GRAVITO – ELECTRIC POWER GENERATION

Gravito-electric power generation is similar to hydro-electric power generation in which the energy of water falling in Earth's gravitational field powers water-turbines that drive electric generators.

In gravito-electric power, Figure 12-1 below, a gravitation deflector makes the water in the central region of the mechanism lighter than that in the outer region, which is acted on by natural gravitation. The lighter reduced gravitation water floats up on the in-flow under it of the heavier natural gravitation water. The result is continuous circulation of the water, like a continuous waterfall.

Water turbines as used in hydro-electric plants placed in the gravito-electric continuous water flow drive electric generators as in hydro-electric plants.

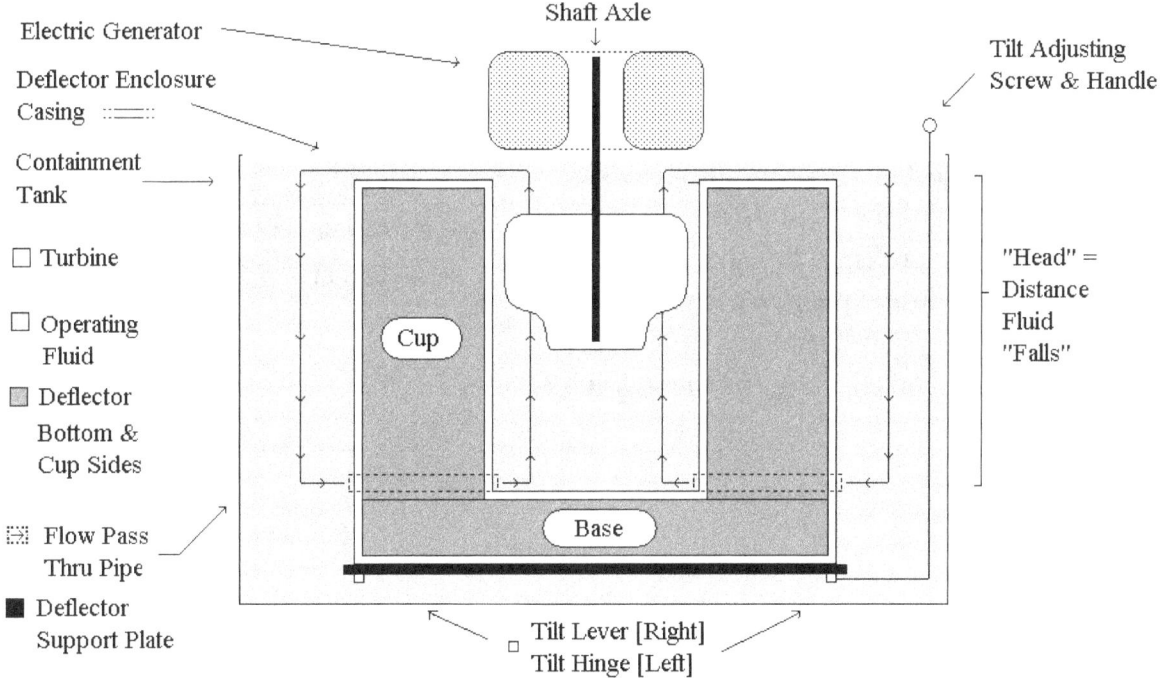

Notes: - The deflector height varies per desired "head".
        - The operating fluid is water or hydraulic fluid.

Figure 12-1
Gravito-Electric Power Generation

203

Deflector Design - Page 1 of 3

Base Top View

Cup Top View

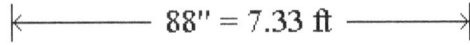 = four ingots: round, square, or hexagonal [shown 4" square]

|← — — 88" = 7.33 ft — — →|

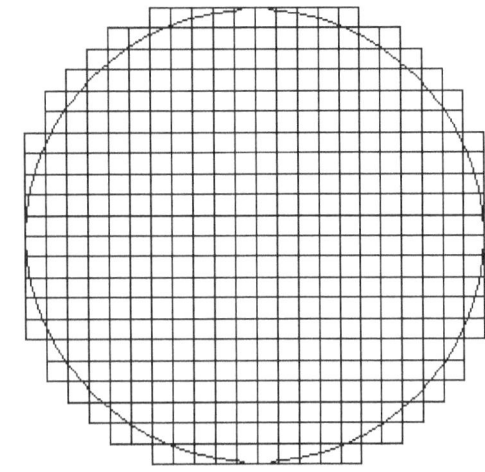

| side | space | side |

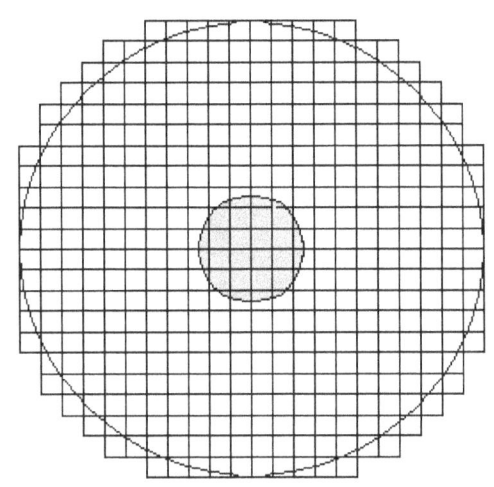

Ingot Design:
- Round is rejected - does not nest closely, leaves open spaces.
- Hexagonal is preferred to square because hexagonal is less expensive per page 3.
- Hexagonal crystals "nest" in a fashion that is more firm and stable as shown on page 3.

side = cup side =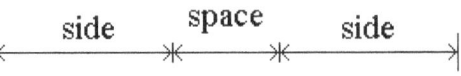
thickness = 85 cm = 34"
space = open center of cup =
diameter = 50 cm = 20"

Deflector Design - Page 2 of 3

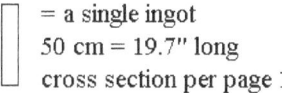 = a single ingot
50 cm = 19.7" long
cross section per page 1

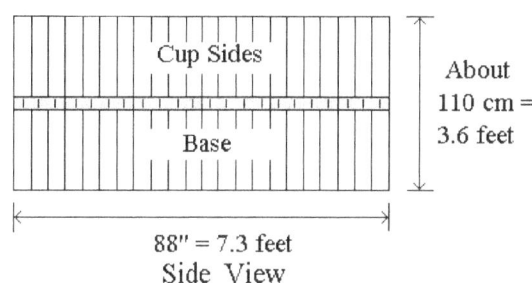 = pipe for fluid
circulation passage
under, and support
for, cup sides about
10 cm OD

Cup Sides

Base

About
110 cm =
3.6 feet

88" = 7.3 feet
Side View

Ingots Count for Hexagonal Ingots

Base = one layer = 641 ingots [416 of page 1 diagram × 1.54 per page 3]
Side = one layer = 616 ingots [416 - 16 = 400 of page 1 × 1.54 per page 3]

1,257

The weight of one such ingot = 16.56 pounds
The total weight of the deflector = 20,816 pounds

GRAVITATIONAL APPLICATIONS

Deflector Design - Page 3 of 3

Areas

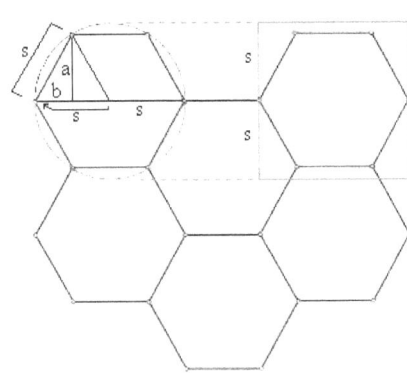

Circle = πs^2
      = $3.14\, s^2$

Square = $[2 \cdot s]^2$
      = $4 \cdot s^2$

Hexagon = 6 equilateral triangles
      = $6 \cdot [½\, a \cdot b] \cdot 2$

           $a = s \cdot Sin\, 60° = s \cdot \dfrac{\sqrt{3}}{2} = 0.866 \cdot s$

           $b = ½ \cdot s$

      = $6 \cdot [½ \cdot 0.866 \cdot s] \cdot [½ \cdot s] \cdot 2$

      = $2.6 \cdot s^2$

"Bare" ingots, ingots as grown, are naturally approximately round.

1. A 4" square ingot [s = 2"] requires a 6" round ingot before machining [for s = 2 the diagonal of the square shown is 5.66, greater than 5].
2. The hexagon shown can be machined from a 4" round ingot.
3. The area of that hexagon is 65% [2.6÷4] of the area of a 4" square.
4. Therefor the number of hexagonal crystals that is required is 1.54 [1÷0.65] times the number of 4" square crystals that would be required (machined from 6" round ingots).

206

$\longrightarrow$

$\longrightarrow$

The Anti-Gravitational Effect

THE DEFLECTION CAUSES A REACTION BACK ON THE DEFLECTOR

Everything in nature is balanced. Nature exhibits a general law of conservation that goes far beyond conservation of energy. For example:

- All positive charge is ultimately, somewhere, balanced by an equal amount of negative charge;

- Gravitational attraction takes place by a mass acting on another mass. The attractive force acting on each is the same in magnitude and opposite in direction; the forces balance;

- The "Big Bang" produced equal amounts of matter and anti-matter;

- For every force there is an equal-but-opposite reaction force;

- Every North magnetic pole is matched by an equal strength South pole.

As that balance, there is a reaction on the deflection-causing gravitation deflector, a reaction to its deflecting action, a balancing reaction.

The gravitational field *Flow*, the *Propagated Outward Flow,* is an essentially unlimited capacity to produce acceleration. That is what the outward propagating gravitational field *Flow* does: it accelerates any and every encountered particle of mass no matter how many and no matter where located.

But, the amount of gravitational acceleration does not depend on the mass that is accelerated; rather, it is in an amount dependent only on the mass, M, of the gravitational *Flow* source and the distance, d, from that source to the accelerated mass, which two parameters determine the gravitational field strength at the accelerated mass.

(13-3) Gravitational Acceleration $= G \cdot M/_{d}2$

That *Flow* is what the gravitational deflector deflects.

The associated "force" is that acceleration multiplied by the mass that is accelerated, which can be whatever mass it happens to be. Thus for gravitation the "force" is inconsequential. No "force" is actually there except in our mental concept of the action. It is the acceleration that is the action.

The deflector reduces the gravitational attraction on all that is <u>above</u> it. 100% deflection is deflecting every "ray" of incoming gravitational *Flow* from its vertical to horizontal, 90° or more, 100% deflection reduces attraction on all that is <u>above</u> it to zero.

The deflection process occurs throughout the length of each <u>deflector crystal</u>. Some rays of gravitational *Flow* are deflected by the first row of atoms of the deflector. Others are deflected by the second row, others the third, and so on. The total deflection is essentially spread linearly uniformly over all of the length of the deflecting crystal. Therefore 100% deflection would reduce the gravitational downward attraction on the <u>deflector itself</u> by only 50%.

On the other hand, the nature of the repulsive reaction is such that 100% deflection means 100% reaction.

The reaction on the deflector is an "equal but opposite" <u>acceleration of the deflector mechanism away from the source</u> of the before deflection gravitational field *Flow*. That is, it acts in the direction opposite from the toward-the-source direction of the acceleration that undeflected gravitation produces. The deflector experiences that reaction acceleration regardless of the mass of the deflector and no matter what additional mass may be attached to it, which attached mass is accelerated with the deflector.

That is because, again, gravitational field *Flow* accelerates any and every encountered particle of mass no matter how many and no matter where located, in amount independent of the mass accelerated, the amount dependent only on the gravitational field strength at the encountered mass.

The ultimate result of the deflection action is the combination of reducing the gravitational attractive acceleration of the deflector [and whatever is attached to it] toward the gravitation source plus the introducing of a reactive repulsive acceleration of the deflector [and whatever is attached to it] in the direction away from the gravitation source.

For example, a deflector that experiences a natural gravitational acceleration, A, reduced by *100%* gravitation deflection to *0.5·A*, plus simultaneously experiencing the reaction to the *100%* deflection in the amount *1.0·A*, experiences a net acceleration acting in the direction <u>away</u> from the gravitation source of *1.0·A − 0.5·A = 0.5·A*.

THE MECHANISM OF THE ANTI-GRAVITATIONAL ACCELERATION

One cannot simply rely on that everything in nature is balanced to account for so dramatic an effect as the repulsive acceleration reaction to the deflection of gravitation – an actual anti-gravity. However, the mechanism producing the effect is simple and natural.

First

Natural gravitational acceleration is caused by that the *Propagated Outward Flow* from a *Spherical-Center-of-Oscillation* encountering another such *Spherical-Center-of-Oscillation* increases the μ and ε concentration on the encountered side of the encountered center. That reduces the encountered center's *Flow* propagation in the direction of that increased μ and ε concentration.

The incoming flow from a distant "source" particle having the effect of slowing the speed of the "encountered" particle's outward propagated flow causes that "encountered" particle's outward flow to have less momentum than if it were not slowed.

Therefore the Newton's Third Law reaction to that reduced outward flow momentum, reaction back on the "encountered" particle, is smaller than otherwise. That effect takes place on the side of the "encountered particle" facing toward the "source" particle from which the slowing - causing flow came.

But, on the opposite side of the "encountered" particle no such slowing of its outward propagated flow is present so that the outward flow there has the full natural momentum and the Newton's Third Law reaction on the particle on that side is the full natural amount.

Consequently, the "encountered" particle experiencing its usual full momentum reaction back on itself on its side opposite that facing the incoming flow from the "source" but experiencing reduced reaction back on itself on its side facing the incoming flow from the "source", experiences a net momentum reaction toward the "source" particle from which the slowing-causing flow came.

Thus the particle experiences momentum increase toward the "source" gravitationally attracting particle which is gravitational attraction.

Second

In the case of the deflector, the components of the incoming vertical gravitational field *Flow* that are curved away from the vertical by the deflector's atom's own *Flow*, by virtue of that deflection, <u>are directed over the top of the atom</u> <u>*opposite from* the bottom side facing the source of the gravitation</u> as depicted schematically in Figure 13-13, repeated below from above.

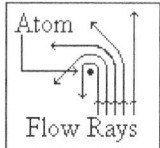

Figure 13-13
Single Atom Deflection of Rays of Gravitational Flow

That increases the *Flow's* μ and ε concentration on that top side of the atom. Just as with natural gravitation, that has the effect of reducing the encountered center's propagation in that direction, the vertically upward direction that of the increased *Flow* concentration caused by the deflected flow.

$\longrightarrow$

SECTION 14

Anti-Gravitation Deep Space Drive

A GRAVITATION DEFLECTOR SPACECRAFT DEEP SPACE DRIVE

A spacecraft gravitation deflector drive would be a deflector in cup form, mounted on the rear of the spacecraft and extending the spacecraft's full length to the nose, as in Figure 14-19 below, with engineered arrangements for varying the amount of deflection [crystal tilt angle].

Figure 14-1
A Gravitation Deflector Driven Spacecraft

This configuration would satisfy a number of functions. The deflector would provide [all without use of fuel]:

- Launching of the spacecraft vertically upward at an upward acceleration of approximately one-half of local natural gravitation, for Earth an acceleration of about 16.1 $ft/_s2$;

- Landing and re-launching of the spacecraft at any gravitating body such as the Moon or Mars;

- Deep space transit propulsion between gravitating bodies;

- Protection from deep space radiation and cosmic ray particles by virtue of the ½ to 1 meter thickness of the Silicon deflector;

- A gravity environment within the spacecraft of zero natural gravitation plus an artificial gravitation due to the acceleration of the ship in whatever amount that it is at any particular time [taking "down" as toward the deflector end of the ship].

213

The engineered arrangements for varying the amount of deflection so as to vary the acceleration would be means of controlled changing of the orientation of selected portions of the Silicon cubic crystals so that they fail to provide the comprehensive deflection of all incoming vertical rays of *Flow*. The engineered arrangements for varying the direction or orientation of the spacecraft would be a 3-axis system of angular momentum wheels

For a spaceship in free space the gravitational *Flow* environment is different from on Earth. In the case of only one gravitation source near enough to be of any important effect and that sole source at a considerable distance from the spaceship, the gravitational *Flow* from that source to the spaceship is essentially all parallel rays. Departing such a source after launch from it requires simply aiming the stern of the ship toward that source. Controlled landing on it requires simply aiming the stern of the ship toward that source and controlling the acceleration by varying the deflection.

In general, however, in deep inter-planetary space gravitation is present albeit fairly weekly because of inverse square reduction of intensity, and it is present in various amounts with attraction toward various differently located sources. As with sailing navigation using the wind as in earlier centuries, spaceship travel within the Solar System may require techniques analogous to: sail craft's tacking on various headings, "crabbing" into partial "cross wind" as aircraft do, and in general going "where the winds permit". In the spacecraft case the "winds" are the various direction gravitational *Flows* available from which to generate acceleration and to which the spacecraft is subject to attraction.

Solar System navigation is further complicated by the destination's continuous motion. The navigation must be toward where the destination will be upon spacecraft arrival at it as compared to where the destination currently is.

For inter-stellar navigation there is the possibility of near light speed travel. The deflector could provide continuous, fuel-less acceleration to the spacecraft throughout its trip. The continuous acceleration would accelerate the craft during the first part and, with the craft re-oriented using the 3-axis system of angular momentum wheels, decelerate the craft for approach to the destination.

Because the acceleration is independent of the mass of the spacecraft it could be quite large and able to carry everything needed for an extended trip and for survival at the destination. The relatively narrow form of the spacecraft is chosen in Figure 14-1 because it provides better shielding against deep space radiation and cosmic rays. A different shape might be chosen for a quite large spacecraft: a single storey flat disk or a wide multi-storey cylinder.

$\longrightarrow$

215

$\longrightarrow$

SECTION 15

Anti-Gravitation Planet Over-Surface Flyer

A GRAVITATION DEFLECTOR PLANET OVER-SURFACE FLYING VEHICLE

A gravitation deflector flying vehicle would be a deflector in cup form, underneath the payload compartment of the vehicle as in Figure 15-1 below.

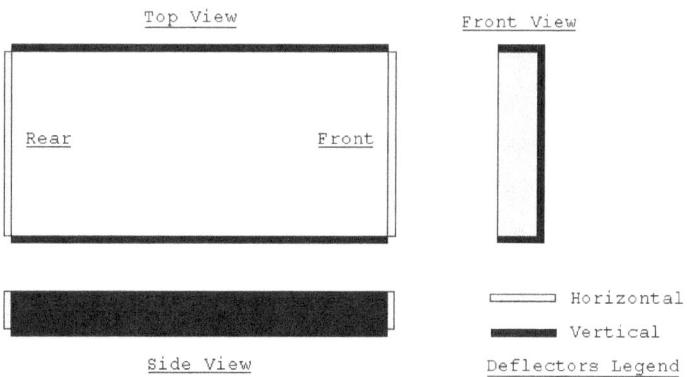

Figure 15-1
A Gravitation Deflector Flying Vehicle

The flying vehicle differs from the form for a spacecraft in:

- not needing to provide protection from dangerous radiation,

- needing only modest acceleration capability vertically upward beyond sufficient to maintain its constant altitude levitation,

- needing means to generate horizontal acceleration while maintaining vertical levitation.

This deflector configuration [all without use of fuel]:

- Provides controlled vehicle levitation for take-off, landing, and travel,

- Provides controlled horizontal propulsive acceleration and "braking",

- But there is the problem of sufficient gravity for the passengers.

217

The vertical acting deflectors cannot provide artificial gravity by virtue of vertical acceleration because the vertical acceleration is controlled to only maintain levitation at a given altitude except for take-off and landing. However, maintaining levitation requires significantly less than 100% vertical deflection. If, for example, levitation required only 50% vertical deflection then the gravitation within the vehicle would be the remaining undeflected 50% of natural gravitation.

The present task, then, is research and development to better optimize the designs so that practical implementation can begin.

$\longrightarrow$

$\longrightarrow$

Appendix F

Relative Propagated Outward Flow Concentrations

PROPAGATED OUTWARD FLOW CONCENTRATIONS

For gravitational applications purposes the interest is in the potential for slowing of the gravitational *Propagated Outward Flow* from the Earth by some configuration of matter at the Earth's surface. The amount of slowing depends on the relative amounts or concentrations of the opposed *Propagated Outward Flow* streams.

Earth Surface Objects Flow Concentration

The ambient *Propagated Outward Flow* within any type of matter is spherically outward from its source *Spherical-Centers-of-Oscillation*. Considering a single such center the successive instants of propagation can be visualized as nested successive hollow shells. Any such shell can be split into two hemispheres, one selected for analysis Then, the radially outward rays of that hemisphere all have a component, u_{amb}, which will be called "u ambient". That situation is depicted in Figure F-1, below.

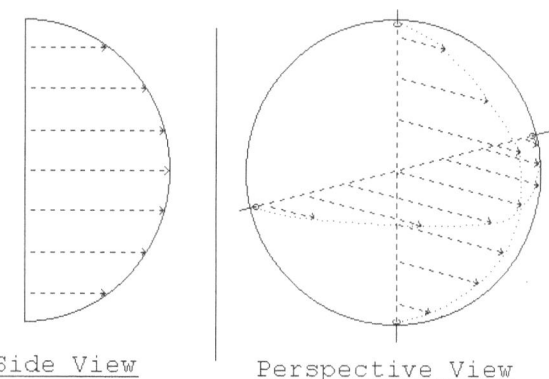

Side View   Perspective View

Figure F-1
Example Rays Comprising u_{amb}

Of course, the rays are not discrete rays neatly arranged along a vertical and a horizontal axis. Rather those shown represent the continuum of medium flow all of the

221

rays of the components of u_{amb}. The average component magnitude corresponds to that hemi-volume divided by the area of the circular base of the hemisphere.

```
(F-1)  r is the radius of the hemisphere, which here
          corresponds to the medium amplitude, u(d),
          where d = r, for a purely radial ray.
```

$$\text{Volume of Hemisphere} = \frac{1}{2} \cdot \frac{4}{3} \cdot \pi \cdot r^3$$

$$\text{Area of Hemisphere Base} = \pi \cdot r^2$$

$$\text{Average } u_2 = \frac{2}{3} \cdot r \text{ and corresponds to } \frac{2}{3} \cdot [u(d=r)]$$

Some example successive stages of the spherically outward *Propagated Outward Flow* from a single center-of-oscillation are depicted in Figure F-2, below.

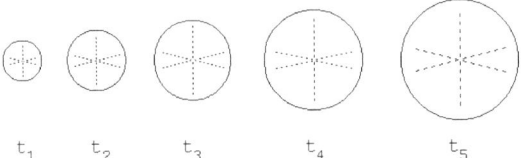

$$t_1 \qquad t_2 \qquad t_3 \qquad t_4 \qquad t_5$$

Figure F-2
Some Stages in a Center's Spherical Propagation

A single stage, such as that of Figure F-1, of the smoothly continuous sequence of stages of which Figure F-2 is a few intermittent examples, is not a solid hemisphere of medium. Rather it is the wave front of medium propagation at an instant of time. A single stage is the outer surface shell of the hemisphere.

The components of medium flow pertaining to that shell act at the curved shell surface, not the theoretical flat circular base of the hemisphere of medium flow. Mathematically one can let the smoothly continuous sequence of such shells be represented by a finite number of nested shells of minute but finite thickness. One such shell is depicted in Figure F-3, below.

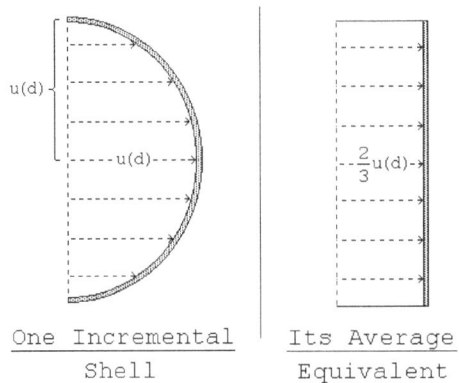

One Incremental
Shell

Its Average
Equivalent

Figure F-3
A Single Theoretical Shell of Medium Flow

The inversF-square variation of the medium flow, $u(d)$, with distance, d, from the center of the source particle from which it is propagated is depicted in Figure F-4, below.

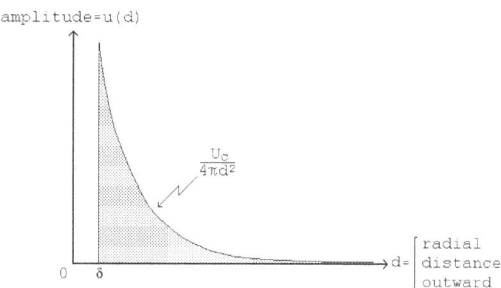

Figure F-4
Propagated Outward Flow Amplitude vs. Distance From Center

This amplitude is actually the concentration, the amount of medium per unit area at the surface of a sphere centered on the *Spherical-Center-of-Oscillation*, as depicted in any single stage of the type depicted in Figure F-2. That amount of medium, itself, is actually the amplitude of the $[1 - Cos]$ form of medium oscillation. [The δ in Figure F-4, above, is the radius of the *Spherical-Center-of-Oscillation*'s core.]

Each atom effectively resides in a cube of side s. The *Spherical-Center-of-Oscillation* of the atom is at the center of the cube and emits *Propagated Outward Flow* in all directions. Per the above Figure F-4, that propagation extends out infinitely in all directions becoming rapidly reduced in magnitude. The cubic volume associated with some single atom experiences the flow of medium from other adjacent and distant atoms through it in addition to its own propagating medium.

Rather than attempt to sum the myriad varied contributions to the medium flow of all of the other affecting sources in the material within a particular atom's volumF-cube, the same net effect can be obtained by attributing all the action of that particular atom (and each individual atom) as taking place within its own volumF-cube. That is, the effect and action per Figure F-4 from $d = \delta$ to ∞ is attributed all to the volumF-cube of its source atom with that volumF-cube unaffected by medium from other atoms.

Assuming a uniform composition of the matter in question, the matter within which the ambient *Propagated Outward Flow* concentration is to be determined, then the average inter-atomic spacing is the same value as the side of the atom's volumF-cube, s. That quantity is the cube root of the reciprocal of the density of the matter times the weight of a single component atom.

The maximum hemisphere centered on the center of the atom, the center of the atom's volumF-cube, as in Figure F-2, that can fit within the cube of volume allotted to the atom is of radius $R = \frac{1}{2} \cdot S$.

The calculation of s is as follows.

(F-2)
$$\text{Density} = \frac{\text{Weight}}{\text{Volume}} = \frac{\text{Atomic Weight}}{S^3}$$

$$S^3 = \frac{1}{\text{Density}} \cdot \text{Atomic Weight}$$

$$= \frac{\text{Total Volume}}{\text{Total Weight}} \cdot \left[\begin{array}{l} \text{Weight of One Atom} = \\ \text{Atomic Mass Number} \times \\ 1.661 \cdot 10^{-27} \ kg/_{amu} \end{array} \right]$$

$$= \text{Volume for One Atom}$$

$$S = [\text{Volume for One Atom}]^{1/3}$$

223

Table F-5, below, gives some typical values for these quantities.

From the table it is clear that inter-atomic spacings, s, in solid elements are on the order of 2.0 to 3.0×10^{-10} *meters*. In a gas at atmospheric pressure the spacing is on the order of 10^{-9} *meters*.

Matter	Density	Weight of Atom	Spacing, S
Air	16	25.9×10^{-27}	1.17×10^{-9}
Water	1000	$18. \times 10^{-27}$	2.62×10^{-10}
Carbon	2250	19.95×10^{-27}	2.07×10^{-10}
Aluminum	2700	44.80×10^{-27}	2.55×10^{-10}
Iron	7870	92.88×10^{-27}	2.28×10^{-10}
Lead	11342	345.35×10^{-27}	3.12×10^{-10}

Table F-5
Some Example Inter-Atomic Spacings

The latest medium flow from the source of u_{amb}, that flow which has not yet propagated outward and inverse square diffused, has the greatest concentration of medium per area, but it intercepts only the smallest area of other source's rays because it is the smallest shell, analogous to $t1$ of Figure F-2. This is the ray of case "a" in Figure F-6, below.

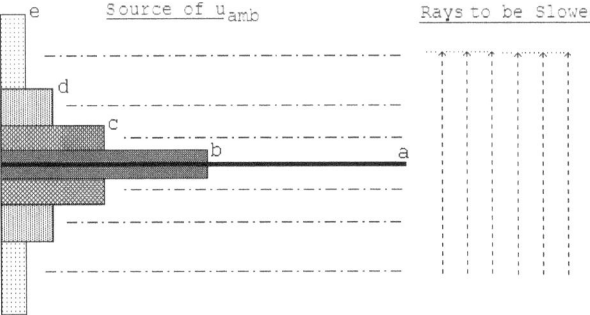

Figure F-6
Encountered Medium Flow for Various Rays

[Note: In Figure F-6 region "e" extends under "d" through "a".
Region "d" extends under "c" through "a". And so forth]

Medium that had been propagated a moment earlier has progressed somewhat in its inverse square diffusion as in case "b" in Figure F-6. Its concentration of medium per area is less because of the distance that it has propagated, but it intercepts a greater area of other source's rays for the same reason. The situation is similar but more progressed with the further successive cases in the figure.

Ray "a" delivers all of the cases depicted in Figure F-6, as indicated in the figure, but it intercepts a smaller area of "rays to be slowed". A source ray that originates at a point some lateral distance away from the center of the source of u_{amb} will encounter only those cases of Figure F-6 which overlay its path but it intercepts a greater area of "rays to be slowed". For example: in the figure ray "c" delivers only cases "a", "b", and "c" however, it intercepts a greater area of "rays to be slowed".

All of the cases from "a" through "e" and beyond, that is all of the shells from $d = \delta$ to $d = \infty$ can be summed as infinitesimally thick individual shells by integration.

An intermediate ray, such as ray "c" in Figure F-6, deliver s all of the cases of shells with a greater radius than the intermediate ray's lateral displacement from the center of the center. Letting r represent that lateral displacement of the ray, D the distance outward from the source of u_{amb} that the shell has traveled, and U_c the fundamental amplitude of the *[1 - Cosine] Spherical-Center-of-Oscillation*, then the summation of the concentrations that that ray encounters in the various shells on outward from lateral displacement r is as follows ($2/3$ is for average per equation F-1.

(F-3)

$$\sum \text{Ray Concentrations} = \frac{2}{3} \int_{r}^{\infty} \frac{U_c}{4\pi \cdot D^2} dD$$

This equation *F-3* is the product of medium flow concentration and a distance (the variable of integration, D). That which is needed is the average medium flow concentration within the atom's volume cube, that is over the range $D = \infty$ to R ($R = \frac{1}{2} \cdot s = \frac{1}{2} \cdot$ *[the volume cube side]*). The integration on the variable D to ∞ then divided by the distance only out to R attributes all of the atom's medium flow propagation solely to its own volumF-cube.

Therefore, dividing equation *F-3* by $[R - \delta] = R$ because $R \gg \delta$ and performing the integration the equation *F-4*, below, is obtained.

(F-4)

$$\sum \text{Ray Concentrations} = \frac{2}{3 \cdot R} \int_{r}^{\infty} \frac{U_c}{4\pi \cdot D^2} dD$$

$$= \frac{U_c}{6\pi \cdot R} \cdot \left[\frac{-1}{D} \right]_{r}^{\infty} = \frac{U_c}{6\pi \cdot R \cdot r}$$

In Figure F-6, while ray #1 encounters the greatest concentration of medium flow, only a very minor portion of the total incoming rays of u_1 can be in position to experience that concentration. On the other hand, ray #2, encounters a reduced medium flow concentration but a much larger number of rays can have that experience. The number of rays that can experience the medium flow concentration for any particular lateral displacement, r, is the area of the concentric ring of radius r and thickness dr. For each of the $r's$ of equation *F-4* the number of ray encountering that concentration is thus $2\pi \cdot r \cdot dr$.

Therefore, equation *F-4*, above, must be integrated by the factor $2\pi \cdot r \cdot dr$ over the range that r can have within the atom's volumF-cube, from $r = \delta$ to $r = R$. That process weights each of the different medium flow concentrations encountered by incoming rays that lie in the successively greater r displacement rings and sums the weighted values. Dividing that result by the overall target area involved, $\pi \cdot [R^2 - \delta^2] = \pi \cdot R^2$ because $R \gg \delta$, gives the average medium flow concentration contributed by actions within the hemisphere of radius R centered on the center of oscillation and oriented toward the flow being encountered.

(F-5)

$$\frac{\text{Average Flow}}{\text{Concentration}} = \frac{1}{\pi \cdot R^2} \int_{\delta}^{R} 2\pi \cdot r \left[\text{Equation E} - 4 \right] d$$

$$= \frac{1}{\pi \cdot R^2} \int_{\delta}^{R} 2\pi \cdot r \left[\frac{U_c}{6\pi \cdot R \cdot r} \right] dr = \frac{1}{\pi \cdot R^2} \int_{\delta}^{R} \left[\frac{U_c}{3 \cdot R} \right] dr$$

$$= \frac{U_c}{3\pi \cdot R^3} \cdot [R - \delta] = \frac{U_c}{3\pi \cdot R^2} \qquad [R \gg \delta]$$

225

This average medium flow concentration contains the only medium flow components, u_{amb}, directly toward the encountered rays present within the hemisphere within the cube of volume allocated to the atom. That medium concentration must be averaged over the overall cube of atomic volume. The result is the average medium flow concentration throughout the hypothesized piece of matter.

(F-6)
$$\frac{\text{Overall Average}}{\text{Concentration}} = \frac{U_c}{3\pi \cdot R^2} \cdot \frac{\text{Hemisphere Volume}}{\text{Atomic Cube Volume}}$$

$$= \frac{U_c}{3\pi \cdot R^2} \cdot \frac{\frac{1}{2}\left[\frac{4}{3} \cdot \pi \cdot R^3\right]}{S^3}$$

$$= \frac{U_c}{9 \cdot S^2} \qquad \left[R = \frac{1}{2} \cdot S\right]$$

However, this calculation has been for a simple *Spherical-Center-of-Oscillation* such as a proton or an electron. The nucleus of an atom is the result of combining A protons and A – Z electrons into one overall new *Spherical-Center-of-Oscillation* oscillating in a complex manner.

The oscillation amplitude is the same for all the various nuclear specie and is not of interest here in that gravitation is an average effect. The average value of the complex oscillation of an atomic nucleus is equal to $z \cdot U_c$. The oscillation [in matter as compared to anti-matter] is entirely within the +U region of medium (with the sole exception of the Hydrogen isotopes, Deuterium and Tritium, which are not of significance here).

That average value is the result, however, of a +U average value of $A \cdot U_c$ and a -U average value of $[A - Z] \cdot U_c$. That is, the atomic nucleus propagates an average medium amplitude of $A \cdot U_c$ in +U and simultaneously a lesser average medium amplitude of $[A - Z] \cdot U_c$ in -U.

Furthermore, the atom's orbital electrons collectively propagate at the same time an average medium amplitude of $z \cdot U_c$ in -U. Those sources of medium flow are not located at the atomic nucleus, but their average effect is as if they were so located because of their orbits around the atomic nucleus.

The total medium flow concentration in a piece of solid matter made up solely of atoms of specie [^{Z}(*Element Symbol*)$_A$] is, then, $A \cdot U_c$ in +U plus $[A - Z] + Z = A \cdot U_c$ in -U. That is a collective medium flow concentration of $2 \cdot A \cdot U_c$. Equation *F-6* then becomes as follows for any such matter.

(F-7)
$$\frac{\text{Medium Flow Concentration}}{\text{Within Matter}} = 2 \cdot [\text{Atomic Mass Number}] \cdot [\text{Equation E} - 6]$$

$$= \frac{2 \cdot A \cdot U_c}{9 \cdot S^2}$$

Using this result, the relative medium flow concentrations in various forms of matter can be compared. This is done at Table F-7, below, for the same substances as listed in the preceding Table F-5, using the values of *S = [the inter-atomic spacing]* from that table.

226

Matter	Atomic Wt, A	Spacing, S	Ambient Medium
Air	14.99 amu	1.17×10^{-9}	$U_c \cdot 2.43 \times 10^{18}$
Water	18.02 "	2.62×10^{-10}	$U_c \cdot 5.83 \times 10^{19}$
Carbon	12.01 "	2.07×10^{-10}	$U_c \cdot 6.23 \times 10^{19}$
Aluminum	26.98 "	2.55×10^{-10}	$U_c \cdot 9.22 \times 10^{19}$
Iron	55.85 "	2.28×10^{-10}	$U_c \cdot 2.39 \times 10^{20}$
Lead	207.19 "	3.12×10^{-10}	$U_c \cdot 4.73 \times 10^{20}$

Table F-7
Some Example Medium Flow Concentrations In Matter

Earth's Gravitational Propagated Outward Flow

Equation *F-7* gives the value of the ambient *Propagated Outward Flow* within matter, which is to selectively slow the incoming gravitational flow.

The gravitational *Propagated Outward Flow* of interest is all the purely vertical components of the overall propagation, all of the horizontal components cancelling each other out to no net effect.

The gravitational acceleration produced by one proton acting on a second proton at a separation distance of one meter is as follows.

(F-8)
$$a_g = G \cdot \frac{m_p}{d^2}$$
$$= (6.67 \cdot 10^{-11}) \cdot \frac{1.67 \cdot 10^{-27}}{1^2}$$
$$= 1.12 \cdot 10^{-37} \text{ meter}/\text{second}^2$$

The medium flow concentration producing that acceleration is as follows.

(F-9)
$$u_g = \frac{U_c}{4\pi \cdot 1^2} = U_c \cdot [7.96 \cdot 10^{-2}]$$

The ratio of these two, that is the gravitational acceleration per amount of medium flow concentration is:

(F-10)
$$\frac{a_g}{u_g} = \frac{1.12 \cdot 10^{-37}}{U_c \cdot [7.96 \cdot 10^{-2}]}$$
$$= \frac{1.41 \cdot 10^{-36}}{U_c} \text{ relative } \text{meter}/\text{second}^2$$

However, this result is only the case when the source of the gravitational field is a proton having a proton's mass, and, therefore, a proton's *Propagated Outward Flow* oscillation frequency. The gravitational effect is directly proportional to the mass of the source of the gravitational field and the frequency of that source's *Propagated Outward Flow* is directly proportional to its mass.

Therefore, in order to apply in general, equation *F-10* must be multiplied by A, the atomic mass in *amu* of the particular gravitational source, divided by *1.07...* the atomic mass in *amu* of a proton, equation *F-11*.

$$(F\text{-}11) \qquad \frac{a_g}{u_g} = \frac{\left[1.41 \cdot 10^{-36}\right] \cdot A}{1.07 \cdot U_c} = \frac{\left[1.32 \cdot 10^{-36}\right] \cdot A}{U_c} \quad \text{Relative } \frac{m}{s^2}$$

The ambient *Propagated Outward Flow* concentration in any particular direction in the several substances listed in the preceding Table F-7 then corresponds to the following gravitational accelerations.

Matter	Atomic Wt, A	Ambient Medium	Grav Accel'n
Air	14.99 amu	$U_c \cdot 2.43 \times 10^{18}$	4.81×10^{-17}
Water	18.02 "	$U_c \cdot 5.83 \times 10^{19}$	1.39×10^{-15}
Carbon	12.01 "	$U_c \cdot 6.23 \times 10^{19}$	9.88×10^{-16}
Aluminum	26.98 "	$U_c \cdot 9.22 \times 10^{19}$	3.28×10^{-15}
Iron	55.85 "	$U_c \cdot 2.39 \times 10^{20}$	1.76×10^{-14}
Lead	207.19 "	$U_c \cdot 4.73 \times 10^{20}$	1.29×10^{-13}

Table F-8
Example Ambient Internal Gravitational Accelerations in Matter

For comparison, the value of the Earth's gravitational acceleration at the surface of the Earth is $9.8 \ m/sec2$.

From Table F-8 the ambient *Propagated Outward Flow* concentrations, available at natural materials' inter-atomic spacings, for producing slowing of incoming gravitational *Propagated Outward Flow* of the Earth are on the order of 10^{15} times too small to have useful effect.

Or, looked at the other way, from equation $F\text{-}10$ the medium flow concentration corresponding to Earth's gravitational acceleration at the surface is

$$(F\text{-}12) \qquad u_g = \frac{[9.8] \cdot U_c}{\left[1.32 \cdot 10^{-36}\right] \cdot A} = \frac{\left[7.94 \cdot 10^{36}\right] \cdot U_c}{A}$$

The principal components of the Earth are approximately as given in Table F-9. From the table the overall average atomic weight, A, of the Earth is about $A = 32.5$.

Earth Component	Percent of Total	Symbol	Atomic Weight	Contribution to Average
Iron	31.0	Fe	55.9	17.3
Oxygen	30.0	O	16.0	4.8
Silicon	16.0	Si	28.1	4.5
Magnesium	15.0	Mg	24.3	3.7
Nickel	2.0	Ni	58.7	1.2
Calcium	1.5	Ca	40.1	0.6
Aluminum	1.3	Al	27.0	0.4
Other	2.0	--	--	--
Earth Average Atomic Weight, *A*				32.5

Table F-9
Earth Average Atomic Weight, A

CONCLUSION AND RATIOS

Therefore, u_g at the Earths' surface is on the order of

228

$$u_{gravitational} \approx 2 \cdot 10^{35} \cdot U_c$$

compared to the ambient U-wave flow concentrations in matter of on the order of

$$u_{ambient} \approx 1 \cdot 10^{20} \cdot U_c$$

per the preceding Table F-8 so that

$$u_{gravitational} \approx 10^{15} \cdot u_{ambient}$$

To use matter at the Earth surface to deflect natural Earth gravitation the effective ambient flow concentration of an Earth surface gravitation deflector must be enhanced by a factor of at least 10^{15}.

$\longrightarrow$

APPENDIX G

Factors Affecting Cubic Crystal Tilt

If it were possible to set the minute tilt angle so that the minute offset of $3 \cdot 10^{-19}$ meter as called for by the Section 4 development could be precisely set and maintained, the "fundamental case", such that the first lattice layer offset is that amount and successive multiples of it sequentially are in the successive layers [2nd layer offset is twice the initial layer; 3rd layer offset is thrice the initial layer; etc.], that direct approach would be taken.

However, the setting of such a minute angle and offset, much less doing so sufficiently precisely, is not practical and probably impossible. To operate using a larger and less precise tilt angle, any tilt angle, the same sufficient number of layers overall required for that "fundamental case" must be employed and the tilt must be such that the actual *x-axis offset* and the actual *y-axis offset* are such that, after that "same sufficient number of layers overall", each required atomic position appears somewhere, in some layer, even though not necessarily in "sequential order".

THE EXACT SUBMULTIPLE OF INTERATOMIC SPACING ISSUE

The most obvious condition of tilt angle and offset that would interfere with "each required effect appearing somewhere, in some layer" would be the actual *offsets* being an exact sub-multiple of the actual interatomic spacing.

For example: with a tilt angle tangent of 0.01, a tilt angle of $0.57°$, the layer-to-layer offset would be 0.01 of an inter-atomic spacing. Layer #2 would be offset $0.01 \times (2.7 \cdot 10^{-10}) = 2.7 \cdot 10^{-12}$ meter from layer #1, layer #3 the same from layer #2 ..., and layer #101 would be offset a total of $2.7 \cdot 10^{-10}$ meter, the actual interatomic spacing, from layer #1.

In that circumstance any further layers would only reproduce the atom locations relative to the vertical U-wave flux that the first 100 layers had introduced.

But if the layer-to-layer offset were such that by layer #101 they accumulated an additional $3 \cdot 10^{-19}$ meter total offset from layer #1, then the second 100 layers would deliver atoms all spaced that $3 \cdot 10^{-19}$ meter beyond the atoms of the corresponding layers of the first 100 .and the third 100 layers would be correspondingly offset from the second, and so on to ultimately delivering an atom in each of $3 \cdot 10^9$ intervals in each $2.7 \cdot 10^{-10}$ meter interatomic space horizontally in the crystal.

If that "additional $3 \cdot 10^{-19}$ *meter* total offset" were, instead, any integer multiple of that amount [but still much less than the $2.7 \cdot 10^{-10}$ actual interatomic spacing] the same overall result would obtain – the in effect shuffling of the layers of the cubic crystal lattice.

The actual *offsets* not being an exact sub-multiple of the actual interatomic spacing is essentially automatically assured. The inverse, requiring a perfect integral sub-multiple relationship would be essentially impossible in practice. That is determined as follows.

A rational number is a number that can be expressed as the ratio of two integers. A rational number expressed as a decimal fraction always exhibits a repetition, over and over, of the sequence of digits in its expression, for example: $0.3333333 \ldots = {}^1/_3$ or $0.125125125 = {}^1/_8$. Conversion of a repeating decimal fraction to the ratio of two integers is done as in *G-1*.

Any number exhibiting such a repeating sequence is rational. Any number that does not exhibit such a repeating sequence is not rational, cannot be converted per equation *G-1*, above, and is therefore irrational.

Consequently, while the number of rational numbers in any interval is finite, the number of irrational numbers in any interval is infinite.

```
(G-1)  [a] The fraction is defined as "F".
           a, b, c, … are digits from the set: 0, 1, … 9
           F = 0.abcd … abcd … abcd …

       [b] Where n = number of digits in F then
           10ⁿ·F = abcd ….abcd … abcd … abcd …

       [c] Then, using n = 4 as an example:
           10⁴·F – F = abcd
           9999·F = abcd

       [d] F = abcd/9999
```

Now consider set *N*, a set of *n* integers [*n* finite]: *1, 2, 3, … , n*. That set is finite, has a finite number of members, is countable and enumerable. Now consider set *R*, all rational numbers such that each such number has a member of *N* as its numerator and a member of *N* as its denominator. There are *n* members of *N*. There are n^2 members of *R*. The number of members of *N* and of *R* is finite.

Now consider set *I*, all irrational numbers greater than *zero* and less than *n*. The number of members of set *I* is infinite. Therefore, the random selection of any number in the interval zero to n, has an infinite probability of being irrational and an infinitesimal chance of being rational.

Therefore, for any installed tilt angle and the offset that it produces, the chance that it would be an exact sub-multiple of the actual interatomic spacing is nil.

On the other hand, the chance that some particular achieved tilt angle and offset requires more layers of cubic crystal than the "fundamental case" because of inefficient scheduling of successive positions is a significant consideration.

TEMPERATURE VARIATION

In addition, a number of variable natural effects are much greater than the precise offset of $3 \cdot 10^{-19}$ meter. The effects of temperature variation in the Silicon cubic crystal and various random vibrations within it would overwhelm such a minute setting.

Most materials tend to expand with increase in their temperature. The measure of that effect is the Thermal Coefficient of Expansion, α. That coefficient relates to thermal expansion of the material as in equation G-2.

(G-2) $\Delta L = \alpha \cdot L \cdot \Delta T$

where:
ΔT = change in temperature in degrees centigrade*
L = length of a dimension of the material
ΔL = change in L due to ΔT
α = thermal coefficient

For Silicon $\alpha = 3 \cdot 10^{-6}$ per degree Kelvin*
at 20° Centigrade

For the interatomic spacing of Silicon the effect of temperature change is per equation G-3, below.

(G-3) $\Delta L = \alpha \cdot L \cdot \Delta T$

$= [3 \cdot 10^{-6}] \times [2.7 \cdot 10^{-10}]$

$= 8 \cdot 10^{-16}$ meters per degree Kelvin

That as compared to the offset of $3 \cdot 10^{-19}$ meter to be created by the tilt. The one-degree temperature variation is over $2,600$ times the objective offset. Even a $1/1000$ degree temperature variation is over double the objective offset. For that reason alone, the setting and maintaining of so precise an objective offset is impractical.

The thermal coefficient of Silicon itself varies with temperature. More precisely it ranges from 2.6 to 3.3 [$\times 10^{-6}$] over the temperature range of $20°$ to $100°$ C.

THERMAL VIBRATIONS AND BLACK BODY RADIATION

The Silicon crystal, at more or less room temperature and as most other materials at that energy, continuously radiates heat energy at frequencies in the infrared range. [The wavelength of infrared radiation is in the range of $3 \cdot 10^{-4}$ to $3 \cdot 10^{-7}$ meters, its frequency being in the range of 10^{12} to 10^{15} Hz [cycles per second].] The crystal also simultaneously absorbs the same kind of radiation from other objects. Its atoms are continuously oscillating, vibrating. Such radiation of that energy comes from a reduction in some atoms' oscillations and such absorption is to an increase in some atoms' vibrations.

If the crystal's temperature is greater than its surroundings its radiated energy exceeds that absorbed and it cools down toward thermal equilibrium with its surroundings. Conversely, if it is cooler than its surroundings its temperature increases due to its absorbing more energy than it radiates.

The heat energy corresponding to the crystals' temperature exists in the crystal as the vibratory oscillations of its atoms about their neutral [*temperature = 0° Kelvin*] position. In a crystal lattice the atoms are bound to their average positions by the neighboring atoms. The spectrum of lattice vibrations ranges from low frequencies to ones on the order of 10^{13} *Hz*.

The dependency of atoms' vibrations on its neighbors depends on temperature. At room temperature range most of the thermal energy is in the vibrations of highest frequency. Because of the short corresponding wavelength the motion of neighboring atoms is essentially uncorrelated so that the vibrations can be considered as independently vibrating atoms, each moving about its average position in three dimensions.

[At high temperature they are not independent of each other. At higher temperatures, not applicable to the present analysis the adjacent atoms are more interrelated in their motions and result in oscillatory waves in the crystal lattice.]

The thermal expansion with increase in temperature is due to the increased amount of energy [heat] in the crystal, and the consequent increase in the amplitude of the crystal's vibrations. The above calculated change in length in Silicon per degree centigrade is the change in amplitude of the atom's vibration.

The *ΔL per degree K* [*= degree C*] of $8 \cdot 10^{-16}$ *meters per interatomic space* of equation *G-2* is about *1 part in* $3 \cdot 10^5$ of the interatomic space. Being the amplitude <u>change</u> that occurs per degree in a range of *100* or more degrees that implies a larger overall amplitude of on the order of *100* or more times that, $8 \cdot 10^{-14}$ *meters*, or about *0.0003* of the interatomic spacing.

The overall net effect of this is that atomic locations and interatomic spacings are continuously shifting and varying in oscillatory fashion. The amplitude of these shifts is only a small fraction of the interatomic spacing of $2.7 \cdot 10^{-10}$ *meters*.

On the other hand the amplitude of those shifts is on the order of over *250,000* times the $3 \cdot 10^{-19}$ *meter* objective distance from an atom that is sought to be achieved. That is, the lattice thermal vibrations cause the atoms to oscillate back and forth about their nominal neutral position a distance on the order of over *250,000* times the $3 \cdot 10^{-19}$ *meter* objective distance from an atom that it is sought to cause all of the Earth's gravitational field to pass in some layer of the crystal.

That $8 \cdot 10^{-14}$ *meters* oscillatory atomic location range covers over *250,000* desirable or suitable objective atomic locations each one of which is effectively randomly sampled or occupied by the thermal vibrations along with all of the others.

THE RANDOM DISTRIBUTION SOLUTION TO THE CRYSTAL TILT

The original concept of the cubic crystal deflector sought to so position the crystal by tilting it relative to the cubic structure of the crystal that atoms of the crystal are forced to effectively occur at successive locations equivalent to a very close, dense positioning of the atoms as seen from the point of view of the purely vertically upward direction of the rays of the Earth's gravitational field. Such positioning would insure that all of the gravitational field is forced to pass extremely close to an atom somewhere in the crystal and to accordingly be deflected away from its natural vertically upward path.

However, the atoms of the Silicon cubic crystal lattice are not fixed in location relative to each other but, rather, are continuously oscillating or vibrating about their nominal neutral positions.

- The vibrations are of various random amplitudes in a range of amplitudes depending on the temperature-determined energy of the vibrations.

- The vibrations are at various random frequencies again in a range of frequencies depending on the temperature.

- The motion of the atoms in their vibration spans a range of locations encompassing a large number of atomic positions that would have been sought to be achieved in various different layers somewhere in the crystal under the original concept and plan.

- That range of motion of the atoms is a small fraction of the neutral interatomic spacing in the crystal.

The net effect of this behavior of the atoms is that the original concept is unworkable. The natural effects of temperature variation and lattice vibrations are much greater than the precise offset intended of $3 \cdot 10^{-19}$ $meter$. The effects would overwhelm such a minute setting, which is probably too minute to accurately set in any case.

The alternative is to accept the random vibratory behavior of the atoms and incorporate it into the overall design.

First, the vibrations of each atom are largely independent of the behavior of the other atoms because the amplitude of each atom's vibrations are such a small fraction, 0.0003, of the interatomic spacing. At any instant of time the totality of the atoms behaving randomly means that, for a sufficient total number of atoms [a sufficiently thick crystal], every sought position of an atom appears somewhere in the crystal. The range of the atom's vibrations can be thought of as a single "super atom" that simultaneously is at all of the $3 \cdot 10^{-19}$ $meter$ intervals in its range.

Second, the issue of the tilt angle and offset that it produces now is that of properly staggering the atomic vibration ranges of the atoms in each layer that same range amount. That is, the tilt objective is now to offset the second layer from the first layer by one atomic vibration range, 0.0003, of the interatomic spacing, $0.0003 \times 2.7 \cdot 10^{-10} = 8 \cdot 10^{-14}$ $meters$.

With the "unit" atomic vibration range being $8 \cdot 10^{-14}$ $meters$ then the tangent of the tilt angle to schedule that range at an equal offset, $8 \cdot 10^{-14}$ $meters$, in each successive adjacent layer is per equation $G-4$.

(G-4)
$$\text{Tan(Tilt)} = \frac{\text{Offset}}{\text{Vertical Layer Thickness}}$$

$$= \frac{8 \cdot 10^{-14}}{5.4 \cdot 10^{-10}} = 0.00015$$

$$\text{Tilt} = 0.008°$$

Figure G-1, below illustrates the effect of equation *G-4*.

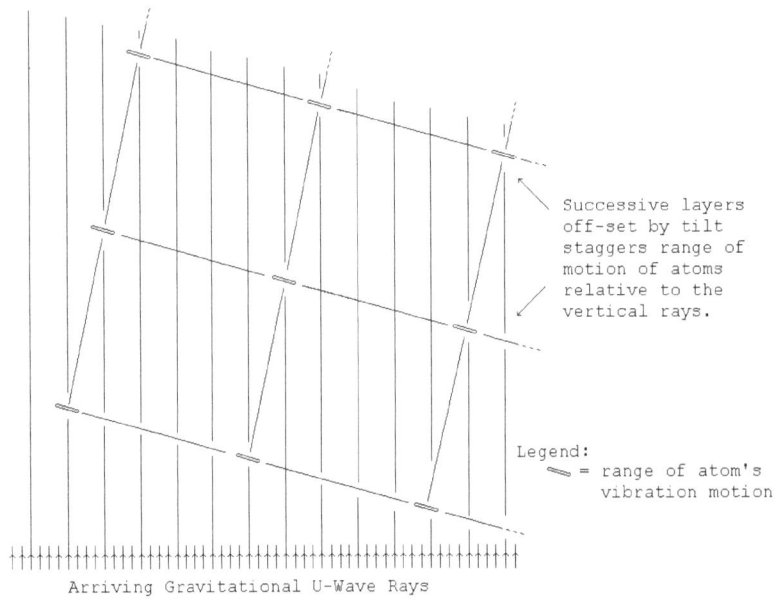

Successive layers
off-set by tilt
staggers range of
motion of atoms
relative to the
vertical rays.

Legend:
≈ = range of atom's
vibration motion

Arriving Gravitational U-Wave Rays

Figure G-1
Cubic Crystal Tilt and Atomic Range of Motion Offset
[Not to scale]

If, instead, the layer to layer offset is set at eleven times the "unit" atomic vibration range of $8 \cdot 10^{-14}$ *meters*, that is *[11]* × *[8·10⁻¹⁴] meters* in each successive adjacent layer the tilt is per equation *G-5*.

(G-5)
$$\text{Tan(Tilt)} = \frac{\text{Offset}}{\text{Vertical Layer Thickness}}$$

$$= \frac{[11] \times [8 \cdot 10^{-14}]}{5.4 \cdot 10^{-10}} = 0.002$$

$$\text{Tilt} = 0.1°$$

That tilt is a not unreasonable value to implement. With it every eleventh layer picks up the position of the second, then third, etc. layer of the equation *G-4* case, the layer to layer offset being equal to the atomic vibration range.

Using a suitably thick section of a commercially grown Silicon cubic crystal ingot *30 cm* in diameter. the equation *G-5* tilt angle tangent of *0.002* would be achieved with a *0.6 mm* thick shim at the edge of the crystal.

236

Going still farther, if, instead, the layer to layer offset is set at one hundred one times the "unit" atomic vibration range of $8 \cdot 10^{-14}$ meters, that is $[101] \times [8 \cdot 10^{-14}]$ meters in each successive adjacent layer the tilt is per equation G-6.

(G-6)
$$\text{Tan(Tilt)} = \frac{\text{Offset}}{\text{Vertical Layer Thickness}}$$

$$= \frac{[101] \times [8 \cdot 10^{-14}]}{5.4 \cdot 10^{-10}} = 0.015$$

Tilt = 0.86°

That almost $1°$ tilt is a reasonable value to implement. With it every $101st$ layer picks up the position of the second, then third, etc. layer of the equation $G-4$ case, the layer to layer offset there being equal to the atomic vibration range.

Using a suitably thick section of a commercially grown Silicon cubic crystal ingot 30 cm in diameter. the equation $G-6$ tilt angle tangent of 0.015 would be achieved with a 4.5 mm thick shim at the edge of the crystal.

PRECISION AND ERRORS

If the intended 4.5 mm thick shim were in error by, for example, about ±0.5 mm then the actual tilt angle tangent would be about 0.013 to 0.017. That corresponds to the multiple of the "unit" atomic vibration range being about 87.75 to about 114.75. Either value or any others in that range will eventually produce all of the desired configurations given sufficient layers.

Another precision issue is that of the orientation of the cubic crystal. For the tilt angle to be precise, the bottom of the cubic crystal slab must be exactly one simple layer of the crystal, that is perfectly aligned to the cubic lattice. In addition the surface on which the crystal and shim rest must be perfectly horizontal.

Furthermore, two shims are needed, one for the x-offset and one for the y-offset. Each must be located at a point on the edge of the crystal corresponding to the midpoint of the interatomic spacing central to the greatest parallel diameter. They must be located at 90° relative to each other, corresponding to two adjacent horizontal sides of the cubic structure.

ANALYSIS OF VARIABLES

There are several quantities or factors bearing on the amount of gravitational deflection produced by a Silicon cubic crystal deflector as here contemplated.

First is the ratio of the U-wave concentration of Earth's natural gravitation as compared to the U-wave concentration in the light diffracted at a slit, calculated at 10^{15} in Appendix B. This ratio determines the closeness required of the passage of gravitational U-waves to atoms of the crystal as calculated in Section 4. This quantity is not variable, but, in spite of Appendix B its value is to some extent an estimate rather than a hard fact. The effect of variation in its value is to correspondingly vary the gravitational   U-wave passage atomic closeness needed, which in the strict "fundamental case" varies the precise tilt angle called for. Practically, that is of no significance in view of the above "The Random Distribution Solution to the Crystal Tilt".

However variation in the gravitational U-wave passage atomic closeness needed produces variation in the number of Silicon cubic crystal layers required which translates into variation in the required thickness of the Silicon cubic crystal slab.

Next is the issue of the vibration range of the crystal's atoms. The range of each of the crystal's atom's vibration varies from that of every other atom because the heat energy so stored varies and continuously changes through exchanges. The atoms' average or typical vibration range is taken above to be about $8 \cdot 10^{-14}$ meters. Those ranges and the uniformity over each range of the random distribution of each individual atom's momentary position in the range are approximate and variable.

The extent to which a particular angle of tilt produces comprehensive coverage of the entire crystal by placing the ranges in successive crystal layers exactly adjacent to each other [as viewed by the vertical flowing gravitational U-waves] or, better, sufficiently overlapping, is a variable because the ranges are a variable. The effect of the degree to which that is optimum or not affects the percentage of the total gravitational U-wave flux that is deflected.

In addition the atomic vibrations are three-dimensional although the analysis has treated only one-dimensional vibrations as in Figure G-1.

Finally, the calculated required thickness of the Silicon cubic crystal slab or ingot, 49 cm varies due to all of the above variations. The effect of this is to affect the percentage of the total gravitational U-wave flux that is deflected, also. In general, the thicker the slab the more deflection likely to be achieved.

PRELIMINARY DESIGN SUMMARY

Pending the results of further research and development experiments the principle design parameters for the initial cubic crystal gravitic deflector are as summarized below.

Per the calculations of Section 4, a silicon monolithic cubic crystal slab 50 cm thick or more should result in 100% deflection.

Common commercially produced silicon cubic crystal wafers are on the order of 600 micro-meters [0.6 mm] thick and up to 30 cm in diameter. Using commercial wafers of that type with their very small thickness would be impractical.

Therefore a single thick slab is needed such as is commercially produced to form the ingot from which the commercial wafers are sliced.

With regard to the distance from the top of the cubic crystal deflector to the bottom of the object above it, the greater that distance is the more effectively reduced is the gravitational U-wave flux acting on the object because the scattered rays of gravitational U-wave can more effectively disperse as they have more distance to travel to the vicinity of the object.

The deflector consists of:

- A support having a verified horizontal upper surface for the cubic crystal deflector to rest upon;

- A Silicon cubic crystal slab:

  · *30 cm* in diameter,

  · *50 cm* or more thick, and

  · with the orientation of the cubic structure determined and noted so that the tilt-producing shims can be properly located at the mid-point of two adjacent sides of the horizontal plane of the cubic structure;

- Precision shims *4.5 mm* thick for producing the tilt of the cubic crystal slab: a tilt angle tangent of *0.015* producing a tilt angle of *0.86°*. on the *30 cm* diameter Silicon cubic crystal slab.

———————   ———————

$\longrightarrow$

$$u(t) = U_c \cdot [1 - \cos(2\pi f t)] \cdot \varepsilon^{-t/\tau}$$

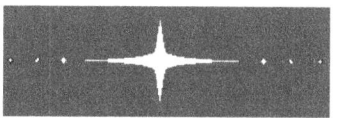